NAVIGATION PRIMER FOR FISHERMEN

SECOND EDITION

W9-ABS-058

NAVIGATION PRIMER FOR FISHERMEN

SECOND EDITION

Captain F. S. Howell, M.B.E.

Fishing News Books Ltd

1 Long Garden Walk, Farnham, Surrey, England

024222

© Captain F. S. Howell, 1977, 1986

First published 1977
Second edition 1986

British Library CIP Data

Howell, F. S.
  Navigation primer for fishermen – 2nd ed.
  1. Fisheries navigation
  I. Title
  623.89′0246392   SH343.8

  ISBN 0 85238 139 5

Published by
Fishing News Books Ltd.
1 Long Garden Walk,
Farnham, Surrey, England

First printed in Great Britain by
Unwin Brothers Limited
Old Woking, Surrey
Second edition printed in Great Britain by
Adlard & Son Ltd.
South Street
Dorking, Surrey

Contents

Chapter 1   The Compass   1

*Magnetic Compass — Magnetism — Variation — Deviation —
Compass Error — Finding Deviation by Transit Bearings and other Shore
Bearings — Deviation Card — Use of Pelorus — Correction of Courses
and Bearings.*

Chapter 2   Charts and Navigational Publications   20

*Introduction — Definitions — Use and Construction of Charts — Chart
Symbols and Abbreviations — Correcting Charts — Notices to
Mariners — Pilots — Admiralty List of Lights — Tidal Information —
Set and Drift of Tides — Merchant Shipping Notices.*

Chapter 3   Chartwork   40

*Use of Instruments — Plotting Positions — Notations on the
Chart — Courses and Distances — Bearings and Cross Bearings —
Running Fixes — Setting a Course to counteract the effects of a
Tide — To find the Set and Drift of a Tide — Speed, Time and
Distance — Leeway.*

Chapter 4   The Sextant   54

*Description of the Sextant — Principles of Construction — Index
Error — Adjustments — Measurement and Correction of Altitudes —
Horizontal and Vertical Sextant Angles.*

Chapter 5   Use of Traverse Tables.   Plane and Mercator Sailing   69

*Difference of Latitude — Difference of Longitude — Departure —
Finding Position by the use of Traverse Tables — Plane Sailing —
Parallel Sailing — Mercator Sailing — Meridional Parts.*

List of Figures

List of Tables

Acknowledgements

I am obliged to the Controller of Her Majesty's Stationery Office, and the Hydrographer of the Navy for permission to reproduce portions of Admiralty charts and publications including *The Nautical Almanac*.

I am indebted to the following for photographs of their equipment and for reproductions from their publications.

Imray Laurie Norie and Wilson Ltd. for extracts from *Norie's Nautical Tables* and *Navigation Primer for Yachtsmen*.

B. Cooke & Son Ltd. for its Kingston micrometer sextant.

Capt F. S. Howell

Preface to the First Edition

The object of this work is to help all fishermen secure an understanding of the basic principles of navigation and so enhance security and win steady success. More specifically it sets out simply and clearly with many illustrations the basic requirements, not only in navigation, but also in chartwork, pilotage and ship stability, for those seeking to qualify for second hand certificate whether full, limited or special.

The author has had many years' experience in teaching fishermen and feels that this book will fulfil a need within the fishing industry for those who are considering obtaining their certificates at a later date. Despite the expansion in electronic equipment in fishing boats within the last few years, the Department of Trade still expects candidates to be able to use and understand the basic skills of navigation, and with this the author agrees entirely.

Before considering taking any of the examinations applicants must be sure that their age, and sea-going experience, is in accordance with the examination regulations, and they should obtain a copy of these which are entitled — *Examinations for Certificates of Competency, Skippers and Second Hands of Fishing Boats, Regulations*, which can be obtained from Her Majesty's Stationery Office, or ordered from any bookseller. A brief extract from the requirements of the regulations is at the beginning of this book, but it is essential that applicants obtain the fullest details. Another important matter is to ensure that they hold a valid eyesight test certificate; details of these tests can be obtained at any Mercantile Marine Office.

The contents of this primer deal with the written parts of the examinations, together with some parts of the orals, but it is advisable for applicants to have a good seamanship book, such as the *Trawlermen's Handbook*, published by Fishing News Books Ltd, to cover the remainder of the orals.

My sincere thanks are due to Captain T. G. Nelson for his assistance in checking the text and making a number of valuable suggestions.

CAPTAIN F. S. HOWELL, M.B.E.
Further Education Teacher's Certificate

Preface to the Second Edition

In this, the Second Edition, the opportunity has been taken to update some of the material contained in the First Edition, and to introduce some new sections, including navigating in the Southern Hemisphere so that the Primer can now be used world-wide.

Changes have been made in some of the Department of Trade Examination Regulations for Certificates of Competancy, and others are expected to follow within the next few years. Candidates for such Certificates are recommended strongly to apply to the Department of Trade for an up-to-date copy of these Regulations which can be obtained from any of Her Majesty's Stationery Offices, or ordered through a bookseller. Consultation with a Nautical College is essential.

A chapter on modern electronics navigation aids, available to smaller craft such as fishing vessels is now included.

CAPTAIN F. S. HOWELL, M.B.E.
Further Education Teacher's Certificate

Chapter 1   The Compass

The Magnetic Compass

The magnetic compass is used to indicate the direction of compass North, and is contained within a bowl made of brass, copper, or plastic material, which in turn is fitted into a binnacle, or attached to a bracket.

To a graduated compass card are attached parallel watertight tubular cases containing magnetised needles—a magnetised ring is used in modern compasses. The card and needles (or ring) are secured to a hemispherical float, or pivot cap, which rests upon a hardened frictionless pin centred in the bowl.

To keep the compass card horizontal and prevent it tilting the centre of gravity of the whole weight of the card, needles (or ring), and float is kept well below the pivoting point as shown in the following diagram.

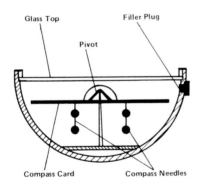

Fig 1   Section diagram of magnetic compass

To stop the compass card swinging violently, due to the motion of the boat, the movement of the card is dampened by the bowl being filled with a liquid consisting of distilled water and part pure alcohol (this is to stop the liquid freezing in extremely cold weather). The compass bowl is also slung in gimbals which further aids it to remain in a more or less horizontal position.

If the liquid in the compass bowl evaporates or leaks out a bubble will appear beneath the glass top. To eliminate this turn the bowl on its side with the filler plug uppermost, unscrew the plug and refill with distilled water. Move the bowl gently from side to side until the bubble disappears and then replace the plug, making sure that the rubber or leather washer on the plug has not worn or perished, otherwise a further leakage will occur.

A wire or painted lubber's line inside the bowl (Fig 2) indicates the direction of the boat's head, and great care must be taken to ensure that this is lined up correctly with the fore and aft line of the boat when the compass is installed. In some cases the compass bowl is suspended in the binnacle, and in others it is secured to a bracket.

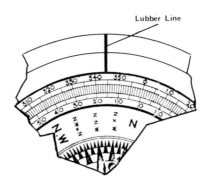

Fig 2   Lubber line

It is essential that the compass be sited so as to be easily visible to the helmsman, but, if only one compass is carried and is used also for taking bearings, it should be sited so that the horizon is visible over a wide arc (it would be expecting too much for a clear vision to be available around the whole 360°). For a compass sited overhead within the wheelhouse it might be necessary to head the boat at a light or shore object to obtain an accurate bearing.

Graduation of Compass Cards

The compass card is usually graduated from 0° at the north point clockwise in degrees to 360°; this is known as the three figure notation because, when using this type of card, the course, or bearing, is expressed in three figures, *eg* 020°, 115°, 278° *etc.*

Older compass cards may be graduated in the quadrantal method which uses the cardinal points (north, south, east and west) and degrees to describe a direction. The four quadrants are each divided into 90°, with zero at north and south, and 90° at east and west. For example, we would express a course or bearing as N 28° E, S 72°E, or N 58° W, and so on.

The points method is rarely used now, but the compass card is divided into 32 points each of 11¼°. Each point is subdivided into half and quarter points. Points are still used to describe the direction of the wind, for example, one would say the wind was SW and not S 45° W or 225°.

The graduated compass card (Fig 3) indicates the three different methods of graduating a card.

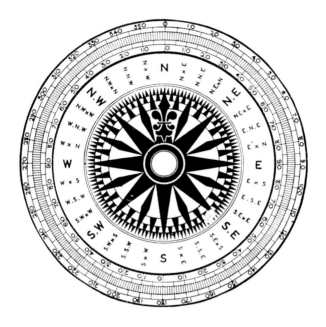

Fig 3 Graduated compass card showing methods of graduation

Magnetism

A profound knowledge of magnetism is not required by candidates for the Second Hand Certificates, but a basic understanding of the principles involved and the effects upon a compass are necessary.

First of all one must know that a bar magnet has two 'magnetic poles' near its extremities. Of these 'poles' the one which sets towards north when the magnet is freely suspended is conventionally termed the 'red pole', and the other the 'blue pole'. One law of magnetism is that like poles repel one another and unlike poles attract each other.

In the area of a magnet there is a zone of magnetic influence known as the 'magnetic field' which may be represented, as in Fig 4, by lines of magnetic force emerging from the 'red pole' and terminating in the 'blue

3

pole'. Other magnets within this magnetic field will align themselves in the direction of the lines of force.

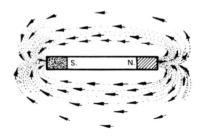

Fig 4   Bar magnet and lines of magnetic force: North pole — red, South — blue

The Earth itself possesses a magnetic field with the 'blue pole' towards the north and a 'red pole' towards the south, and it is this magnetic field which imparts directive force to the magnetic compass. The Earth as a magnet may be likened to a sphere with a small but immensely strong magnet near its centre.

The 'red' or north seeking end of the compass needle will be attracted to the 'blue pole' of the Earth situated in the north, and which is known as the North Magnetic Pole. Fig 5 shows how the compass needle in the northern hemisphere is attracted to the North Magnetic Pole from various positions.

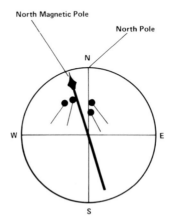

Fig 5   Effect of north magnetic pole on compass needles in the northern hemisphere

One other important point that must be learned is that as the magnetic poles are situated deep within the centre of the Earth, the magnetic lines of force enter the Earth's surface perpendicularly at these poles and are horizontal at the magnetic equator midway between the magnetic poles.

The effect upon a compass needle (except at the magnetic equator) is to cause it to 'dip' towards the magnetic pole and in the seas around the United Kingdom the 'dip' is as much as $67\frac{1}{2}°$. Hence the need to construct the compass card with the centre of gravity well below the pivoting point which prevents this 'dip' or tilt as described previously.

Variation

The Earth rotates on its axis and the points where the axis meets the surface of the globe are called the 'true poles' of the Earth, or Geographical Poles. The north and south magnetic poles are situated at some distance from the True poles, and their respective positions are Latitude 70° North, Longitude 97° West, and Latitude 72° South, Longitude 154° East.

Joining the True poles are the True meridians, and between the magnetic poles are the magnetic meridians. Variation is defined as the angle contained between the true and magnetic meridians. As the magnetic poles are slowly changing their positions each year, variation changes annually, and it is important that the correct variation for the current year should be used.

Information concerning the amount of variation at any particular place is obtained from the chart, either from a 'compass rose' as shown in Fig 6, or from 'isogonic' lines which are drawn on a chart joining all places with the same variation. Special charts are printed which are called *Magnetic Variation Charts* and they show the variation over large sea areas by means of these isogonic lines.

The outer compass is true and the inner compass magnetic, so that the angle between true north and magnetic north at that position is 9° 30′ West in that year (1964).

As a practical test study a chart in current use at sea now, and work out the correct variation for the current year based on Fig 7 (see next page).

Remember that variation is the same no matter in which direction the boat is heading.

Deviation

Deviation of the compass is one of the most elusive factors in the whole practice of navigation. Iron and steel are the metals principally affected by magnetism, and each part of the boat, fitting, or electrical equipment, can exercise a different magnetic effect upon the compass.

If the boat had no magnetic influences, such as ferrous metals in the

5

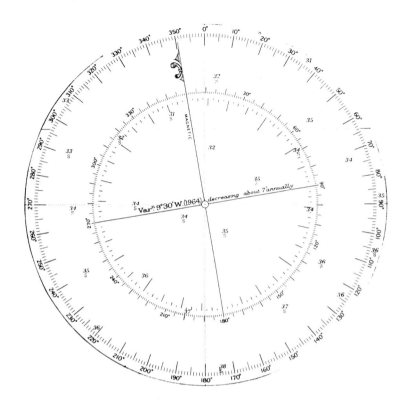

Fig 6   Compass rose showing variation of 9°30′ west 1964 and decreasing about 7′ annually

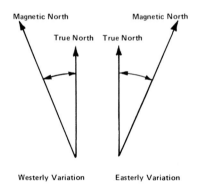

Fig 7   Westerly and easterly variation from true north

construction or fittings, the compass needle would point to magnetic north.

The deflection of the compass needle from magnetic north is called 'deviation', and Fig 8 shows how it can affect the compass. Its study will help understanding of why deviation depends upon the direction of the ship's head.

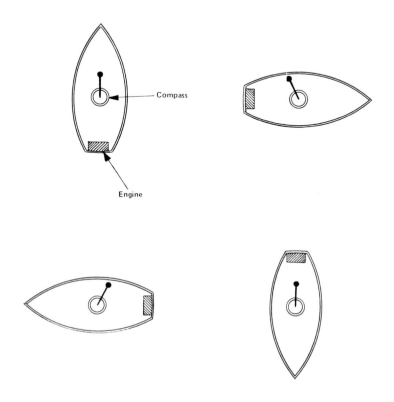

Fig. 8 Deviation depends on direction of ship's head and magnetic influences of vessel and gear

Assuming that the engine in this boat produces a total magnetic field with a blue pole, when it is heading north or south, the compass needle is in line with the blue pole and is not deflected. When heading east the red point of the compass needle is attracted to the blue polarity of the engine (remember unlike poles attract), and this gives a westerly deviation. When heading west the compass needle is again attracted to the blue polarity of the engine thus giving an easterly deviation.

Assume now that there is a trawl winch forward of the wheelhouse and that this winch has a blue polarity, it will be seen that the deviation would be opposite to the previous example. Again there would be no deflection

7

when heading north or south. And when heading east, the red point of the compass needle would be attracted to the blue polarity thus giving an easterly deviation. Equally when heading west, a westerly deviation is caused. Fig 9 illustrates this point.

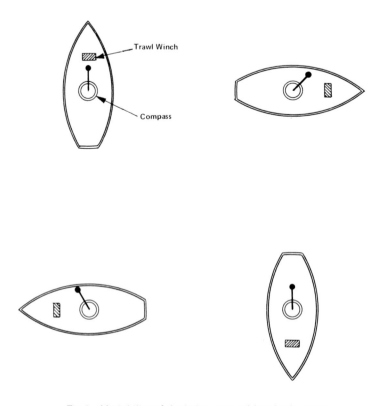

Fig 9 Variability of deviation caused by ship's gear

Correction for deviation is essential to be able to steer accurate courses or take reliable bearings.

Compass Error
When variation and deviation are combined, the result is the error of the compass, and this is the angle between true north and compass north.

For example, with a variation of 10° West and a deviation of 5° West the compass error would be 15° West. If, on the other hand, the variation was 10° West and the deviation 5° East the compass error would be only 5° West.

The simple rule to find compass error is to add variation and deviation when they are of the same name. When they are of different names subtract them calling the error the same name as the larger of the two. Error, variation and deviation are named east or west—east when the north point of the compass is drawn to the right, west when the north point is drawn to the left. Study Fig 10.

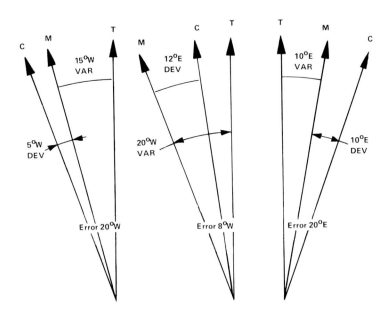

Fig 10   Calculation of Compass Error

Finding Deviation by Transit and Shore Bearings

It is essential that the compass is checked for deviation from time to time, and the practical methods of doing this are easy to learn.

First of all, by means of a transit bearing, look for two conspicuous objects on shore that are marked on the chart, *eg* church, flagstaff, lighthouse, tower, beacon, *etc*. Draw a line through these two objects and read off the true bearing (from seaward). To this bearing apply the variation, adding if west and subtracting if east, and this will give the magnetic bearing. Position the boat on this bearing and take a compass bearing of the two objects in transit (in line). The difference between the magnetic bearing and the compass bearing will be the deviation.

By taking a series of compass bearings with the boat on different

9

headings from north around 360° in either points, or at 10° intervals, one will be able to make up an individual deviation card such as that described below.

If two suitable objects in line cannot be found, look for a single conspicuous object ashore about a mile distant, and with the boat's own position accurately fixed, draw a line from the shore object to the boat's position. This will permit finding the true bearing of the object and then by applying variation one will have the magnetic bearing. A compass bearing of the object taken on different headings, as in the previous example, will give the difference between compass and magnetic bearings, and thus the deviation. It is essential that when one swings the boat it be done around a buoy, or at anchor, so that the relative position with the shore object does not change.

A most useful piece of doggerel which will help to remember the application of variation, deviation and error is:

Variation, deviation or error west compass *best*

Variation, deviation or error east compass *least*

It is so important to remember this that it is repeated later in the chapter

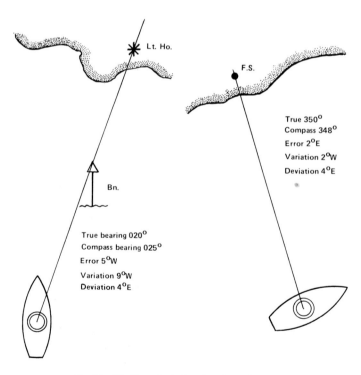

Fig 11   Finding deviation by shore bearings

Another method of finding the magnetic bearing of a single object ashore, when not certain of one's own position, is to take the compass bearing of the object when the boat is heading on the eight cardinal points (or even more if a more accurate result is desired; see Fig 11). As in the previous case the boat must be swung within its relative position to the shore object.

Here is a simple example of a boat taking the compass bearing of a shore object when swung through the eight cardinal points.

Boat's head	Compass bearing	Magnetic bearing	Deviation
North	023°	025°	2° E
NE	026°	025°	1° W
East	029°	025°	4° W
SE	031°	025°	6° W
South	028°	025°	3° W
SW	024°	025°	1° E
West	020°	025°	5° E
NW	019°	025°	6° E

Total 8/200

025° Magnetic bearing

When the boat is new, and at regular intervals thereafter, one should have the compass checked (or swung) for deviation. The compass adjuster doing this will be able to remove any substantial amounts of deviation that may exist, but any deviation remaining would be detailed on a deviation card, which should be displayed prominently in the wheelhouse.

Deviation Cards

Deviation Cards can vary and four different types are shown here Nos. 1, 2, 3 and 4 some of which will be used for exercises later in the book.

Previously Deviation Cards were shown with the Boat's Head only in points of the compass (North, N×E, NNE, etc.) but with 11¼° between each point this led to unnecessary complications when changing from points into degrees as shown on Cards Nos. 1 and 2 (page 12).

If one is required to find the deviation for a compass heading, which is not given exactly in the tables, one must learn to interpolate, or in other words, solve the problem by simple proportion. Practice in this will be needed to enable an applicant to work confidently at the examination.

As an example, using Deviation Card No. 1, what would be the deviation for a compass heading of 131°? For 124° the deviation is 8° W, and for 135° it is 5° W, so that for a change of 11° (that is 135° minus 124°), the change of deviation is 3° (8° W minus 5° W). What is required is the

11

Dev. Card No. 1

Compass Heading	Deviation	Magnetic Heading
000°	4° W	356°
011¼°	6° W	005¼°
022½°	8° W	014½°
033¾°	10° W	023¾°
045°	12° W	033°
056¼°	14° W	042¼°
067½°	15° W	052½°
078¾°	16° W	062¾°
090°	15° W	075°
101¼°	13° W	088¼°
112½°	11° W	101½°
123¾°	8° W	115¾°
135°	5° W	130°
146¼°	3° W	143¼°
157½°	0°	157½°
168¾°	3° E	171¾°
180°	6° E	186°
191¼°	9° E	200¼°
202½°	12° E	214½°
213¾°	13° E	226¾°
225°	14° E	239°
236¼°	14° E	250¼°
247½°	13° E	260½°
258¾°	12° E	270¾°
270°	10° E	280°
281¼°	9° E	290¼°
292½°	7° E	299½°
303¾°	3° E	306¾°
315°	1° E	316°
326¼°	1° W	325¼°
337½°	3° W	334½°
348¾°	4° W	344¾°
000°	4° W	356°

Dev. Card No. 2

Compass Heading	Deviation	Magnetic Heading
000°	16° E	016°
011¼°	13° E	024¼°
022½°	9° E	031½°
033¾°	4° E	037¾°
045°	0°	045°
056¼°	4° W	052¼°
067½°	7° W	060½°
078¾°	11° W	067¾°
090°	14° W	076°
101¼°	18° W	083¼°
112½°	21° W	091½°
123¾°	23° W	100¾°
135°	24° W	111°
146¼°	23° W	123¼°
157½°	21° W	136½°
168¾°	18° W	150¾°
180°	14° W	166°
191¼°	10° W	181¼°
202½°	7° W	195½°
213¾°	3° W	210¾°
225°	0°	225°
236¼°	3° E	239¼°
247½°	7° E	254½°
258¾°	10° E	268¾°
270°	13° E	283°
281¼°	17° E	298¼°
292½°	20° E	312½°
303¾°	22° E	325¾°
315°	23° E	338°
326¼°	22° E	348¼°
337½°	21° E	358½°
348¾°	19° E	007¾°
000°	16° E	016°

deviation for 131° which is 4° from 135°. The simple proportion sum now resolves itself in the following manner:

As 11° is to 3° so is 4° to x (the difference we require). Three degrees multiplied by four degrees divided by eleven degrees, or

$$\frac{3° \times 4°}{11°} \text{ equals } \frac{12°}{11°} \text{ equals } 1° \text{ (to the nearest degree).}$$

Therefore if the deviation for 135° is 5° W, for 131° it must be 5° plus 1° equals 6° W.

12

<table>
<tr><td colspan="3" align="center">Dev. Card No. 3</td></tr>
</table>

Ship's Head by Compass	Deviation	Magnetic Heading
000°	4° W	356°
010°	6° W	004°
020°	8° W	012°
030°	10° W	020°
040°	12° W	028°
050°	13° W	037°
060°	15° W	045°
070°	17° W	053°
080°	18.5° W	061.5°
090°	19° W	071°
100°	20° W	080°
110°	18.5° W	091.5°
120°	17° W	103°
130°	15° W	115°
140°	12° W	128°
150°	9° W	141°
160°	7° W	153°
170°	5° W	165°
180°	2° W	178°
190°	Nil	190°
200°	2° E	202°
210°	5° E	215°
220°	7° E	227°
230°	10° E	240°
240°	12° E	252°
250°	13° E	263°
260°	14° E	274°
270°	12° E	282°
280°	9° E	289°
290°	6° E	296°
300°	4° E	304°
310°	2° E	312°
320°	1° E	321°
330°	1° W	329°
340°	3° W	337°
350°	3.5° W	346.5°

Dev. Card No. 4

Ship's Head by Compass	Deviation	Magnetic Heading
000°	4° W	
020°	8° W	
040°	12° W	
060°	15° W	
080°	18.5° W	
100°	20° W	
120°	17° W	
140°	12° W	128°
160°	7° W	
180°	2° W	
200°	2° E	
220°	7° E	
240°	12° E	
260°	14° E	
280°	9° E	289°
300°	4° E	
320°	1° E	
340°	3° W	

The modern Deviation Cards now give the deviation for every 10° as on Card No. 3, but at the examinations for Certificates of Competency it is now customary to give the deviation for every 20° as on Card No. 4, thereby increasing the amounts of interpolation required, the Column 'Magnetic Heading' is left blank and this must be completed by you when working your chart problems.

To find the deviation for a given compass heading Card No. 3.

(a) Given a compass heading of 137° the deviation for 130° is 15° W and for 140° it is 12° W so by interpolation the sum resolves itself in the following manner.

Between 130° and 140° is 10° and between 15° W and 12° W is 3°, and the difference beween 140° and the compass heading of 137° is 3°. Therefore, as 10° is to 3° so is 3° to x (the difference required) or (3°×3°)/10° equals 9°/10° equals 1° (nearly).

If the deviation for 140° is 12° then for 137° it is 12° W plus 1° equals 13° W which would be the correct deviation to use if taking any compass bearings while heading 137° by compass.

To find the deviation for a given magnetic heading Card No. 3.

(b) If a true course of 277° was laid off on a chart with a variation of 7° W taken from the chart, the magnetic course would be 284°. Turning to the column on the Deviation Card No. 3 marked 'Magnetic Heading' the deviation for 282° is 12° E and for 289° it is 9° E.

Between 282° and 289° the change is 7° and between 12° E and 9° E the change of deviation is 3°. The difference between 282° and the magnetic heading of 284° is 2°, therefore as 7° is to 3° so is 2° to x (the difference required), or (3°×2°)/7° equals 6°/7° equals 1° nearly which gives a deviation of 11° E, and applied to the magnetic heading of 284° gives a compass course of 273°.

Using Card No. 4 the working would be as follows: (a) Deviation for compass heading of 137°. Deviation for 120° is 17° W, and for 140° it is 12° W, therefore as 20° is to 5° so is 3° to x, or (5°×3°)/20° equals 15°/20° equals 1° (nearly). Deviation for 140° is 12° W and for 137° it would be 12° W plus 1° equals 13° W.

(b) True course of 277°, variation 7° W equals magnetic course of 284°. For magnetic heading of 274° the deviation is 14° E, and for 289° is 9° E, therefore as 15° is to 5° so is 5° is to x, or (5°×5°)/15° equals 25°/15° equals 2° (nearly).

Deviation for 289° is 9° E and for 284° it would be 9° E plus 2° equals Deviation 11° E. Applied to the magnetic course of 284°, 11° E deviation would make a compass course of 273°.

Use of Pelorus

The pelorus (Fig 12) is usually a portable instrument additional to the compass, but in some boats a pelorus may be fixed in a suitable position giving more or less all round visibility. It is a circular metal plate, graduated like a compass card and mounted on a vertical axis so that it can be rotated freely by hand and can be clamped into any position required. It has a lubber line marked on the plate and sight vanes (front sight and back sight).

The plate is turned by hand so that the lubber line indicates the same direction as the boat's compass, and it is used mainly to obtain bearings of objects that may not be visible from the position of the boat's compass and it would also save turning the boat's head to the object as described earlier.

In taking a bearing by pelorus the sight vane is moved until the object is visible through the back sight aperture and is in line with the Fore sight cord. The bearing is read from a pointer at the foot of the Fore sight.

In setting up a pelorus on board one must be sure that its 0–180° line is accurately placed in or parallel to the fore and aft line of the boat just as is done with a compass. When taking the bearing of an object with the pelorus it is set up with the boat's compass course clamped to its lubber line. The helmsman calls out when the boat is steady on the compass course, and the bearing is then observed through the sight vanes and read off as if taken directly from a compass. Note that it requires two men to operate, one at the helm and the other at the pelorus.

Another method of using the pelorus is to line it up with 0° as the boat's head and at the moment of sighting an object through the sight vanes, the helmsman calls out the compass heading of the boat, and the bearing of the object would be read off the pelorus as a 'relative' bearing. If, for example, the compass heading was 020° and the relative bearing was 072° then the compass bearing of the object would be 020° plus 072° equalling 092°. If the compass heading and the relative bearing come to over 360° then subtract 360° from it. For example, with a compass heading of 349°

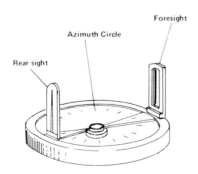

Fig 12   Drawing of a pelorus

024222

and a relative bearing of 320° it would equal 669° minus 360° and so equalling 309°.

A problem sometimes given in examination is to use the pelorus to put the boat on a pre-determined true course from a known position. To do this one would obtain the true bearing of a conspicuous landmark from the boat's known position, and set and clamp the sight vane on to this true bearing. Rotate both the dial and sight vanes until the desired course is lined up with the lubber line. 'Con' the boat until the sight vane points to the land mark when the boat will then be on her true course.

At the same time the helmsman will call out the compass heading and the difference between this and the true course will be the compass error. By applying variation to this compass error one will be able to find the deviation (for that particular compass heading only).

Correction of Courses and Bearings

Having learned the meanings of variation, deviation, compass error, and the use of the deviation card, one comes to their practical application in the correction of courses and bearings.

The following little doggerel will help remembrance of which way to apply compass error.

> Error west, compass *best*.
> Error east, compass *least*.

After one has had practice in applying these corrections they will be found quite simple to understand. As an example, look at the compass rose (Fig 6) and assuming that the total error of the compass in this case is $9\frac{1}{2}°$ W, one should lay the ruler through the centre of the compass rose to where the outer (true) ring is marked 090°. It will be seen immediately that it cuts the inner (magnetic) ring where it is marked $099\frac{1}{2}°$ (error west compass best). This means that if one had a compass error of $9\frac{1}{2}°$ west one would have to steer $099\frac{1}{2}°$ to make good a true course of 090°.

A few worked examples of changing true courses to compass courses will make one familiar with the procedure:

a. True course to steer 230°   Variation 9° W
        Error   7° W   Deviation 2° E
    Compass course to
        steer   237°   Error   7° W

(Error west compass best)

16

b. True course to steer 162°   Variation 2° W
       Error   2° E   Deviation 4° E

      Compass course to
         steer   160°   Error   2° E

      (Error east compass least)

c. True course to steer 358°   Variation 7° W
       Error   4° W   Deviation 3° E

      Compass course to
         steer   002°   Error   4° W

      (Error west compass best)

The next problem to be dealt with is to change Compass bearings into True bearings:

Compass bearing	062°	Compass bearing	002°
Error	6° E	Error	6° W
True bearing	068°	True bearing	356°
(Error east compass least)		(Error west compass best)	

In the above examples the variation has been given, and in practice one would take this from the chart being used—(at examinations it is given in the problem). Deviation also has been given, but in practice (and in the examinations) one will be required to find the deviation from the deviation card. To obtain this it is necessary to know the direction of the boat's head, and this will always be given in the problem.

   Given a true course, and asked to find the compass course to steer, first apply the variation to the true course and this will give the magnetic course; then go to the deviation card and under magnetic heading take out the corresponding deviation, which, when applied to the magnetic course will give the compass course.

   Another useful mnemonic with which to memorise how to apply these corrections is as follows:

                  Compass   = Cadbury's
                  Deviation = Dairy
                  Magnetic = Milk
                  Variation = Very
                  True   = Tasty

One must learn that in going from compass towards true one applies first deviation, and then variation (these two are usually combined to make compass error and are applied as one figure). In going from true towards compass first apply the variation to the true course to obtain the magnetic course, and under magnetic heading take out the required deviation.

If one had a true course of 068° and a variation of 7° W this would give a magnetic heading of 075°. It will be seen from Deviation Card No. 1 that the deviation for this magnetic heading is 15° W, so the compass course would be 090° (Error west compass best).

In practice (and in the examinations) more often than not one finds the magnetic heading comes somewhere between the figures on the deviation card, and interpolation is required as explained previously, for example:

$$\begin{array}{ll} \text{True Course} & 265° \\ \text{Variation} & \underline{\quad 8°} \text{ W} \\ \text{Magnetic course} & \underline{273°} \end{array}$$

From Deviation Card No. 2, for 269° the deviation is 10° E, and for 283° it is 13° E. Therefore, 283° minus 269° equals 14°, and 13° minus 10° equals 3°.

As 273° is 4° from 269° the sum would appear as follows:

As 14° is to 3° 4° is to x

$$\frac{3° \times 4°}{14°} \text{ equals } \frac{12°}{14°} \text{ equals } 1° \text{ (to the nearest degree)}$$

Deviation for 269° is 10° E, and for 273° it is 10° plus 1° equals 11° E.

$$\begin{array}{ll} \text{Magnetic course is} & 273° \\ \text{Deviation} & \underline{\quad 11°} \text{ E} \\ \text{Compass course} & \underline{262°} \end{array}$$

When one is required to correct compass bearings, and is given the boat's head by compass, one can go straight to the deviation card and take out the deviation.

In an examination a problem could be given in the following manner: From a vessel steering 315° C (which means by compass) Start Point Light bore 022° C, and Bolt Head bore 342° C. Find the position of the vessel.

18

Variation 10° W, and using Deviation Card No. 1.

Looking at this deviation card it is seen that the deviation for a compass heading of 315° is 1° E, and with variation 10° W and deviation 1° E the compass error is 9° W.

Put down the two compass bearings and apply the compass error to them, which will give the true bearings.

Start Point Light 022° C   Bolt Head 342° C
Error   9° W   Error   9° W

013° T (True)   333° T (True)

One must fix firmly in the mind the fact that deviation depends upon the direction of the boat's head, and as long as it remains on that heading the deviation is the same. When one applies deviation to a compass heading or bearing one gets the magnetic heading or bearing, and if one applies deviation to a magnetic heading or bearing one gets the compass heading or bearing.

Another useful term to use is the word CADET. Taking the two *outside* letters of CADET and using the three middle letters you can say:

From compass to true *add east*
Obviously you must *subtract* west

Chapter 2   Charts and Navigational Publications

Introduction

Navigation is the science of finding a vessel's position at sea, and of navigating safely from one position to another. A knowledge of charts, *Sailing Directions* (known as 'Pilots'), *Light Lists*, and other publications is required for one to navigate in coastal waters. Additional aids, such as buoys, lighthouses, lights, fog signals, radio beacons, etc. all must be known.

So that these aids can be used, their positions on the Earth's surface must be known, and it is necessary to know something about the Earth, the methods adopted to define positions on the surface of the Earth, and the units of measurement used.

In navigation, outstanding landmarks, soundings of the sea bottom, *etc.* are used to give assistance in finding one's position. Further one will notice on charts that while the landward portion is usually practically blank, except near the coastline, the seaward portion contains many figures, letters, and lines. It is this part that is so important and about which one must learn.

Definitions

The Earth may be represented on a small scale, as a globe with the oceans and continents mapped on its surface. In order to find the exact position of any place on the Earth it is covered with a network of lines, both vertical and horizontal, from which the location of a place can be described.

Latitude

The lines representing latitude are drawn horizontally and parallel to the Equator, and consist of a series of small circles (a small circle is one whose plane does not pass through the centre of the Earth). It will be seen from the drawing below how they diminish in size (circumference) as they approach the poles.

Longitude

The lines which run from pole to pole over the surface of the Earth are called meridians, and they are all great circles because their plane passes through the centre of the Earth. The meridian passing through Greenwich is called the Prime Meridian, or Greenwich Meridian, and from it all Longitude is measured (see Fig 13).

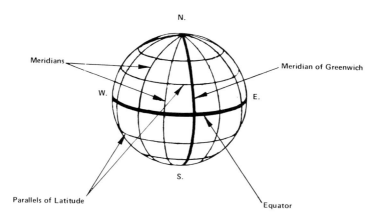

Fig 13 Parallels of latitude and meridians

Position

The position of a place is described in terms of latitude and longitude. For latitude it will be given as so many degrees, minutes and seconds, north or south of the Equator. For longitude the amount in degrees, minutes and seconds is given either east or west of the Greenwich Meridian. For example, the Lizard lighthouse is in 49° 57′ 36″ N 05° 12′ 00″ W.

Angular Measurement of Latitude

This measurement is the angular distance of a parallel of latitude measured along the arc of a meridian north or south of the Equator, and is equal to the same angle drawn from the centre of the earth above or below the Equator, to where a line meets the surface of the earth. For example, an angle of 30° drawn from the centre of the earth above or below the Equator would meet the surface of the earth in latitude 30° N or 30° S.

21

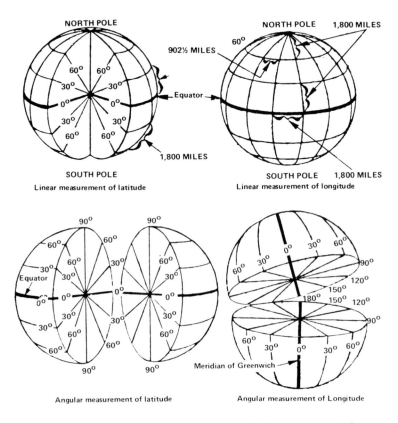

Fig 14 Linear and angular measurement of latitude and longitude

Linear Measurement of Latitude

One minute of latitude equals one nautical mile, which is taken as 6080 feet, therefore, 900′ of latitude on the globe would equal 900 miles in whatever latitude one was situated.

Angular Measurement of Longitude

This measurement is the angular distance east or west of the prime meridian (Greenwich) measured along the Equator to the meridian passing through the position whose longitude it is required to know. Also, it is equal to the angle subtended at the centre of the earth between the two same meridians.

Linear Measurement of Longitude

It will be seen that the distance on the surface of the earth, between any two meridians, diminishes uniformly from the Equator until it is nothing at the north and south poles; therefore, the linear distance of a degree of longitude on the surface of the earth varies with its latitude and cannot be taken as a standard measurement of length (see Fig 14). For example, while 30° of longitude at the Equator is equal to 1,800 nautical miles, in latitude 60° North it is just over 900 nautical miles.

Mercator's Projection

It would be impossible to navigate accurately with the sole aid of a globe, because all courses and bearings would have to be made on a curved surface.

At the end of the 16th century a Dutchman, called Mercator, devised a method whereby a specified area of the globe could be transferred to a flat surface chart. With the object of preserving the uniformity of the chart and keeping all the angles relatively correct, the successive parallels of latitude are drawn at a steadily increasing distance apart, in exactly the same ratio as a degree of longitude is increased at each parallel.

Instead of curving towards the poles, the meridians are drawn as straight lines at right angles to the parallels of latitude, as will be seen on the Mercator chart. As the latitude increases towards the poles, the parallels of latitude increase their distance apart. The distance between two points on a chart is always measured from the latitude scale at the sides of the chart, and as each minute of latitude gets longer towards the poles it is most important that measurements are taken only on that part of the latitude scale which lies abreast of the two points whose distance apart is required (see Fig 15).

Never use the longitude scale to measure distance.

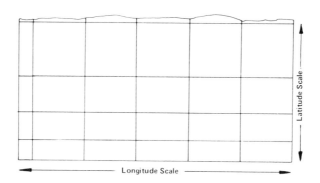

Fig 15   Mercator's projection of latitude and longitude scales

Gnomonic Projections

Some small harbour plans and charts are drawn on the gnomonic projection as are charts for the polar regions.

On a gnomonic chart meridians appear as straight lines converging to the poles, and the parallels of latitude are shown as curves. On a large scale chart or plan covering a small area, the curvature of the parallels and the convergence of the meridians are so slight that the chart may be used in exactly the same manner as any other chart.

On this type of chart distance can be measured either by using the latitude scale or from a separate scale that is marked on the side or bottom of the chart.

Examples of chart scales are shown in Fig 16.

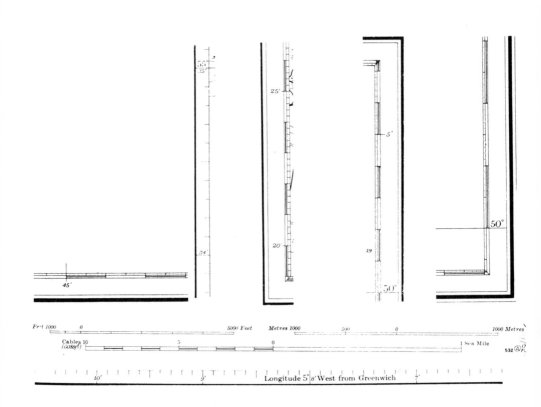

Fig 16   Examples of chart scales

Before proceeding further, study this brief summary of the terms and definitions used so far and memorise them thoroughly.

24

Great circle
Is a circle on a sphere whose plane passes through the sphere's centre.

Small circle
Is a circle on a sphere whose plane does not pass through the centre.

Equator
Is a great circle on the terrestrial sphere midway between the poles.

Poles
The poles of the Earth are the points where the Earth's axis meets the surface.

Parallel of latitude
Is a small circle on the terrestrial sphere, parallel to the equator.

Meridian of longitude
Is a great circle on the terrestrial sphere passing through the poles and cutting the equator at right angles.

Prime meridian
The meridian through Greenwich Observatory (Long. 0°). It is the meridian from which all longitude is measured.

Plane
Is any flat surface.

Nautical mile
Is the length of a minute of arc measured along a meridian, in other words, from your latitude scale.
   The standard length of a nautical mile is 6080 feet.

Cable
Is one-tenth of a nautical mile. In practice it is taken as 600 feet, or 200 yards, or 100 fathoms.

Knot
Is a unit of speed equal to one nautical mile per hour.
   The expression 'knots per hour' should never be used to denote speed, one would say 9 knots, for example.

Chart datum
The level below which soundings are given on Admiralty charts, and above which are given the drying heights in feet of rocks, mud or sand banks, *etc.*, which are periodically covered and uncovered by the tides. It is also the level above which tidal levels and predictions are given in the A.T.T. (*Admiralty Tide table*).
   Chart datum level is the lowest to which the tide normally falls at that particular point, and used to be slightly below Mean Low Water Springs,

but new charts are now based at a lower level called the Lowest Astronomical Tide (L.A.T.) which is the lowest predictable tide under average meteorological conditions.

Classification of Charts

Charts may be divided into four main categories.

World charts
These charts are on a very small scale showing ocean routes, magnetic variations, telegraph cables, ocean currents, etc.

Ocean charts
These charts are on a larger scale covering wide areas of oceans, etc.

Generally they do not show much detail of the coast lines and are used for ocean voyages, but not when making a landfall. Referred to as small scale charts.

General charts
These charts are used over limited areas such as sections of a coastline.

Drawn on a conveniently large scale they are essentially for use in coastal navigation. Referred to as large scale charts.

Plan charts
These charts are on the largest scale and give in great detail harbours and narrow channels, etc.

When coasting the largest scale chart available should always be used.

Publication of Charts

The Hydrographic Department of the Navy Department, at the head of which is the Hydrographer of the Navy, is responsible for the production, printing, and issue of all Admiralty charts and books.

To obtain details about the latest charts published for a particular area one must consult the most recent copy of the *Catalogue of Admiralty Charts*, which can be seen at any Admiralty chart agent or depot for the sale of charts and hydrographic publications, situated at most fishing ports. When ordering a new chart, the title of the chart and also the number should be quoted. These are printed in the bottom right hand corner and the top left hand corner of each chart. Whenever a new chart, or a new edition or reprint is published a notice is given in the *Admiralty Notices to Mariners*, and all old copies should be cancelled and withdrawn from chart agencies and depots upon receipt of the new copies.

To make certain one is obtaining an up to date chart with all the latest information on it one should look at the age and the state of the chart as follows:

1. Date of publication of a new chart is shown in the middle of the outer lower margin of the chart, *eg*

Published, London 30th Oct. 1964 *under the Superintendence of* Rear Admiral E. G. Irving, C.B., O.B.E., *Hydrographer of the Navy.*

2. When a chart is revised throughout, or modernised in style a new edition is published and recorded to the right of the original date of publication, *eg*

New Editions 24th Sept. 1965

3. When large corrections have to be made a reprint is issued containing these corrections, and is recorded below the date of edition, *eg*

New Editions 29th July 1932. 8th Dec. 1933, 10th March 1939, 30th Dec. 1960.

Large Corrections 29th May 1964, 23rd Dec. 1966.

4. Small corrections published in the *Admiralty Notices to Mariners* are made by hand either by the chart agency or depot, or the actual user of the chart. Such small corrections are noted in the bottom left hand corner of the chart by inserting the number of the relevant notice against the year of issue, *eg*

Small corrections 1974 – 16 – 214 etc

When a chart is subsequently reprinted these small corrections are included in the new chart.

Small corrections-1965-2110-2251-1966-261-1967-158-1377-1968-1935.

Chart Symbols and Abbreviations

For a full study of the details contained on charts a copy of *Chart 5011* (issued in book form) entitled *Symbols and Abbreviations used on Admiralty Charts* and published by the Hydrographic Department of the Navy must be obtained.

This is a most comprehensive publication and is divided into various sections covering The Coast-Line, Coastal Features, Topography—Natural Features, Control Points, Units, Adjectives, *etc.*, Ports and Harbours, Topography—Artificial Features, Buildings, Miscellaneous Stations, Lights, Buoys and Beacons, Radio and Radar, Fog Signals, Dangers, Various Limits, Soundings, Depth Contours, Quality of the Bottom, Tides and Currents, The Compass, and Abbreviations of Principal Foreign Terms.

The International Hydrographic Organisation (IHO) publishes a standard list of charted features. The list assigns a letter to each major category of a charted feature (*eg* G denotes Ports and Harbours) and a number to each feature within the category (eg G37—Floating Dock). *Chart 5011* follows the

IHO system. Comparisons between symbols and abbreviations used on Admiralty Charts and corresponding symbols and abbreviations used on the Charts of other member states of the IHO may be made without difficulty. For example, G37 will locate the symbols of a floating dock in this publication and in similar publications issued by most other Hydrographic Offices.

Another most useful complementary publication to assist students in the ability to read and understand charts is *The Mariner's Handbook NP 100* which deals world wide with such subjects as general marine meteorology, tides and currents, navigational hazards, traffic separation systems, *etc*.

Charts are now published in both metric charts and fathoms charts and eventually all charts will become metric, so one should be familiar with both types (see Figs 17 and 18). Metric charts are readily distinguishable from fathoms charts by their improved design and by the greater use of colour. Most symbols and abbreviations used on Admiralty charts are common to metric and fathoms charts, therefore little difficulty will be experienced in transferring from one type to the other, especially as in Chart 5011 the fathoms and metric symbols and abbreviations are printed side by side.

At the examinations candidates will be asked to define certain symbols and abbreviations and to illustrate others, so a thorough study of the contents of *Chart 5011* should be made and the introduction carefully read.

Experience can be gained by studying the symbols and abbreviations on any chart, checking with *Chart 5011* if any cannot be identified.

For exercise, determine the following abbreviations:

Ldg, Submd, Ht, Tr, Fl, OC, AL, RC, R, PA, Wk, M.H.W.S., Sn, P, F.S., C.G., Oy, M. Wd, bk, sm, w, St.

Illustrate the following: eddies, overfalls and tide-rips, rock awash at chart datum, wreck showing any portion of hull at the level of chart datum, breakers, foul, rock which covers and uncovers at 4 feet chart datum.

Check your answers with *Chart 5011*, extracts from two pages of which follow (Figs 17, see opposite page and 18, see pages 30 and 31).

First study the 'Title' of the chart (Fig 19, see page 32).

Note the following points:

1. The area covered by the chart.

2. The date of the latest information.

3. Reference to bearings and the height above chart datum of drying heights and all other heights expressed in feet above mean high water springs.

B Coastal Features

Metric Charts		
I	G.	Gulf
2	B.	Bay
3	Fj.	Fjord
	† Fd	
4	L.	Loch, Lough, Lake
(Ba)	Lag.	Lagoon
	† Lagn	
5	Cr.	Creek
7	Str.	Strait
8	Sd.	Sound
9	Pass.	Passage
10	Chan.	Channel
(Bb)	Appr.	Approaches
	† Apprs	
11	Ent.	Entrance
	† Entce	
(Bc)	R.	River
12	Est.	Estuary
13	Mth.	Mouth
14	Rds.	Roads, Roadstead
15	Anch.	Anchorage
16	Hr.	Harbour
16a	Hn.	Haven
17	P.	Port
18	I.	Island, Islet
19	† It	Islet
20	Arch.	Archipelago
	† Archo	
21	· Pen.	Peninsula
	† Penla	
22	C.	Cape
23	Prom.	Promontory
	† Promy	
24	Hd.	Head, Headland
25	Pt.	Point
26	Mt.	Mountain, Mount
29	Pk.	Peak
30	Vol.	Volcano
33	Lndg.	Landing place
35	Rk.	Rock

† This abbreviation is obsolescent

Fathoms Charts		
I	G.	Gulf
2	B.	Bay
3	F^d	Fjord
4	L.	Loch, Lough, Lake
(Ba)	Lag^n	Lagoon
5	Cr.	Creek
7	Str.	Strait
8	S^d	Sound
9	Pass.	Passage
10	Chan.	Channel
(Bb)	$Appr^s$	Approaches
11	Ent^{ce}	Entrance
(Bc)	R.	River
12	Est^y	Estuary
13	M^{th}	Mouth
14	R^{ds}	Roads, Roadstead
15	$Anch^e$	Anchorage
16	H^r	Harbour
16a	H^n	Haven
17	P.	Port
18	I.	Island, Islet
19	† I^t	Islet
20	$Arch^o$	Archipelago
21	Pen^{la}	Peninsula
22	C.	Cape
23	$Prom^y$	Promontory
24	H^d	Head, Headland
25	P^t	Point
26	M^t	Mountain, Mount
29	Pk	Peak
30	Vol.	Volcano
33	L^{dg}	Landing place
35	R^k	Rock

† This abbreviation is obsolescent

Fig 17   Fathoms and metric charts

Dangers

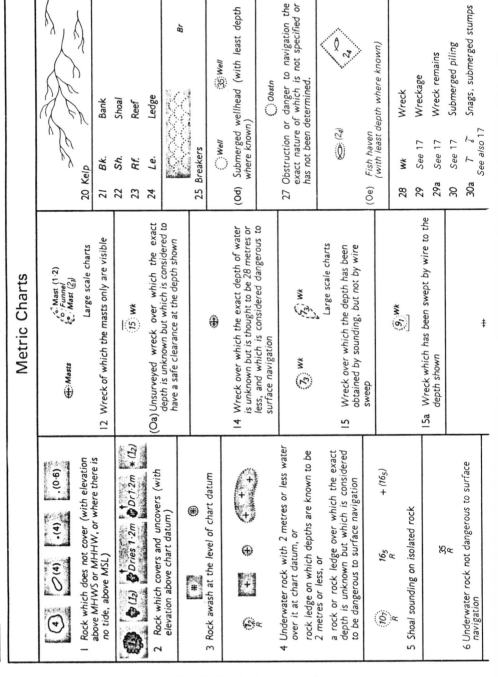

Metric Charts

Masts ⊕
Mast (1·2)
Funnel
Mast (2₂)
Large scale charts

12 Wreck of which the masts only are visible

$\overline{15}$ Wk

(Oa) Unsurveyed wreck over which the exact depth is unknown but which is considered to have a safe clearance at the depth shown

⊕

14 Wreck over which the exact depth of water is unknown but is thought to be 28 metres or less, and which is considered dangerous to surface navigation

(₇₃) Wk   (₇₃) Wk
Large scale charts

15 Wreck over which the depth has been obtained by sounding, but not by wire sweep

(₉) Wk

15a Wreck which has been swept by wire to the depth shown

‡

1 Rock which does not cover (with elevation above MHWS or MHHW, or where there is no tide, above MSL)
(4) (4) ·(4) ·(0·6)

2 Rock which covers and uncovers (with elevation above chart datum)
(1₂) Dries 1·2m Dr 1·2m ★ (1₂)

3 Rock awash at the level of chart datum
⊕

4 Underwater rock with 2 metres or less water over it at chart datum, or
rock ledge on which depths are known to be 2 metres or less, or
a rock or rock ledge over which the exact depth is unknown but which is considered to be dangerous to surface navigation
(2/R) + +

5 Shoal sounding on isolated rock
(10)/R 16₅/R + (16₅)

6 Underwater rock not dangerous to surface navigation
35/R

20 Kelp

21 Bk.   Bank
22 Sh.   Shoal
23 Rf.   Reef
24 Le.   Ledge
Br

25 Breakers

(Od) Well   Well
Submerged wellhead (with least depth where known)

Obstn
27 Obstruction or danger to navigation the exact nature of which is not specified or has not been determined.

(Oe) (2₄)
Fish haven
(with least depth where known)

28 wk   Wreck
29 See 17   Wreckage
29a See 17   Wreck remains
30 See 17   Submerged piling
30a T   Snags, submerged stumps
See also 17

Fig 18   Metric Chart symbols

30

32	dr		Dries
33	cov		Covers
34	uncov		Uncovers
35	Rep		Reported
	†Repd		

38 *Danger line (see Note)*

(Ob) Areas of mobile bottom (including sand waves)

41	PA	†(PA)	Position approximate
42	PD	†(PD)	Position doubtful
43	ED	†(ED)	Existence doubtful
		See Q1	Sounding of doubtful depth
44	pos	†posn	Position
46	unexam	†unexamd	Unexamined

6a Underwater danger with depth cleared by wire drag sweep

Historic Wreck (see Note)

† Historic Wreck (see Note)

(Oc) Restricted area round the site of a wreck of historical and archaeological importance.

(Covers and uncovers)   (Always covered)

10 Coral reef

Large scale charts

11 Wreck showing any portion of hull or super-structure at the level of chart datum

Drying heights: See note in the Introduction.

Non-dangerous wrecks: Where the depth of a wreck exceeds 28 metres, or it is otherwise considered non-dangerous, the appropriate symbol is generally shown on the largest scale chart only.

Danger line: A danger line draws attention to a danger which would not stand out clearly enough if represented solely by its symbol (eg. isolated rock), or delimits an area containing numerous dangers. A danger line through which it is unsafe to navigate. A bold pecked line with explanatory legend may be used to delimit an area where there is inadequate information.

16 Wreck over which the exact depth is unknown but thought to be more than 28 metres, or

a wreck over which the depth is thought to be 28 metres or less, but which is not considered dangerous to surface vessels capable of navigating in the vicinity.

Foul

Foul   Where depth known

17 The remains of a wreck, or other foul area, no longer dangerous to surface navigation, but to be avoided by vessels anchoring, trawling, etc.

18 Overfalls and tide-rips

19 Eddies

† *This symbol and/or abbreviation is obsolescent*

Fig 18   Metric Chart symbols

31

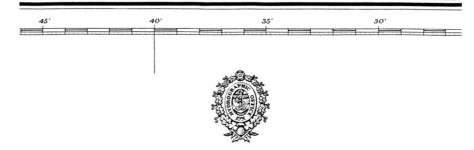

ENGLAND - SOUTH AND WEST COASTS

TREVOSE HEAD

TO

DODMAN POINT

INCLUDING

THE SCILLY ISLES

FROM THE LATEST ADMIRALTY SURVEYS TO 1955.

Soundings in hairline are from smaller scale surveys.
The Topography is taken chiefly from the Ordnance Survey
Bearings refer to the True Compass and are given from Seaward (thus :- 126° etc.)
Underlined figures on drying banks and rocks express the Heights
in Feet above the datum to which the soundings are reduced
All other Heights are expressed in Feet above Mean High Water Springs
For Abbreviations see Admiralty Chart 5011

SOUNDINGS IN FATHOMS
(Under Eleven in Fathoms and Feet)

Natural Scale $\frac{1}{150,000}$ (at Lat. 50°25'N.)

Projection − Mercator

Fig 19   Example of Chart Title in Fathoms

4. Soundings (always check carefully whether these are in fathoms, feet, or metric units).

5. Natural scale—the scale of the chart to the actual distance. For example on this chart it is expressed as $\frac{1}{75,000}$ so that 1 inch on the chart would represent 75,000 inches, which is 6,250 feet, or 1 nautical mile of 6,080 feet, plus 170 feet at the latitude of 50° 10′ N.

6. The projection of the chart. It could be Mercator but may be gnomonic, but would be clearly indicated.

7. Various cautions.

32

ENGLAND — SOUTH COAST

EDDYSTONE ROCKS
TO
BERRY HEAD

DEPTHS in METRES

SCALE 1:75 000 at lat 50°30'

Depths are in metres and are reduced to Chart Datum, which is approximately the level of Lowest Astronomical Tide.

Heights are in metres. Underlined figures are drying heights, in metres and decimetres, above Chart Datum; all other heights are above MHWS.

Projection: Mercator.

Authorities: Mainly from Admiralty Surveys, 1950 to 1968. Soundings south of Eddystone Rocks and west of 3°52'W are from an Admiralty Survey of 1864. The topography is taken chiefly from the Ordnance Survey.

Fig 20   Another example of Chart Title — Eddystone Rocks to Berry Head

Now study the 'Title' of metric chart (*No. 1613*) in Fig 20 above.

In comparing this with a fathoms chart note the reference to the depths being expressed in metres, otherwise the information given is almost identical, except the reference to the chart datum level of approximately the lowest astronomical tide, which has already been described.

Correcting Charts

All small but important corrections affecting navigation that can be made to charts by hand are published in the Weekly editions of the *Admiralty Notices to Mariners*. Always make sure that your charts are kept up to date, and corrections should be made in water-proof violet ink, marking in the number and date of the correction in the bottom left hand corner.

Use the correct abbreviations, shown on *Chart 5011*.

Typical examples of types of chart correction are shown in Figs 21 and 22 overleaf.

Sometimes corrections take the form of a reproduction of a portion of a chart known as 'blocks'. These are cut out from the relevant *Admiralty*

33

*1799. SCOTLAND, W. COAST — Upper Loch Long — Glenmallan — Light altered.

The pier head light in position 56° 07′ 48″ N., 4° 49′ 03″ W. (approx.) is to be amended to *F.G.*

Chart [*Last correction*].—3739 (plan, Approaches to Finnart)—(plan, Loch Long and Loch Goil) [*422/71*].
Light List Vol. A/70, 4413·7.
H.M.S. *Neptune* Hyd. Note 9/70. (*H. 6408/68.*)

Fig 21 Typical example (repro type 1799)

*1319. ENGLAND, S. COAST — WEYMOUTH — The Nothe — D.G. Range Eastward.

(1) A can light-buoy, blue and white chequers, *Qk. Fl. R.*, is to be inserted in position 351° 5·1 cables from "C" Head light (50° 35′ 45″ N., 2° 25′ 55″ W. approx.).

(2) Spar buoys with spherical topmarks are to be inserted in the following positions:—
(a) 262° 120 feet ⎫
(b) 352° 120 feet ⎬ from the light buoy in (1).
(c) 307° 165 feet ⎭

(3) "*D.G. Range*" is to be inserted close W. of the buoys in (2) (*on chart 2255* close W. of the light-buoy in (1)).

Charts [*Last correction*].—2268 [*313/69*]—2255 (1, 3) [*313/69*].
Portland Notice 9/69. (*H. 4340/62.*)

Fig 22 Another example of chart correction (1319)

Notices to Mariners and pasted on to the chart exactly over the portion affected by the corrections, as in Fig 23.

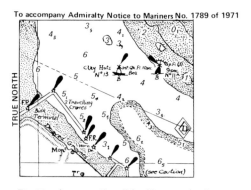

Fig 23 A correction 'block' on a chart

Remember that when correcting charts only the amount of detailed information relevant to that type of chart must be added. For example, do not add a small change of position of a harbour buoy on to an ocean chart which showed that particular harbour on a very small scale.

Admirality Notices to Mariners

Reproduced below is the outside cover of a weekly edition of *Admiralty Notices to Mariners*. Most of the information contained on the cover is self explanatory, but NEMEDRI and CHINPACS might appear as

34

Notices
1873-1931/85.
Aus 348-376/85.
List of T & P Notices in force.
List of Corrections to Sailing Directions.

ADMIRALTY
NOTICES TO MARINERS

WEEKLY EDITION 29

27 July 1985

© Crown Copyright. Permission is not required to make copies of these Notices but such copies are not to be sold

---

CONTENTS

I Index

II Admiralty Notices to Mariners

III Radio Navigational Warnings

IV Corrections to Sailing Directions

V Corrections to Admiralty Lists of Lights and Fog Signals

VI Corrections to Admiralty Lists of Radio Signals

Mariners are requested to inform the Hydrographer of the Navy, Ministry of Defence, Taunton, Somerset TA1 2DN (Telex 46274), immediately of the discovery of new dangers, of changes or defects in aids to navigation and of shortcomings in Admiralty charts or publications. Copies of form H 102, which is a convenient form on which to send in a report, may be obtained *gratis* from any Admiralty Chart Agent or the reproduction at the end of section V of the weekly edition of Notices to Mariners may be used. A copy of the form, which may be used as a *pro forma*, is also printed in The Mariner's Handbook (NP 100).

Taunton   R. O. MORRIS, *Rear-Admiral.*
                             Hydrographer of the Navy.

Fig 24   Title Page of Admiralty Notices to Mariners

strange terms. NEMEDRI gives details of areas dangerous due to mines, swept routes, and buoyage in the North Sea, Baltic, Mediterranean, and Black Sea. CHINPACS gives similar details in the Far East, and the word is a combination of CHina Sea, INdian and PACific oceans.

Fishing boats will not normally carry a large number of charts but care must be taken to ensure these are corrected. The weekly *Notices to Mariners* should be studied closely as a glance at the index will show at once whether any charts possessed are affected, and also whether new editions have been published (see Fig 24).

In addition the annual *Summary of Admiralty Notices to Mariners* should be studied, also the monthly *List of Temporary and Preliminary Notices* in force.

Pilots

In whatever area one is fishing a copy of the *Sailing Directions* covering that particular portion of coastline or sea must be at hand. These amplify the information given on charts and give other general information of interest to the mariner. They are known generally as 'Pilots', and a total of 73 volumes covers the entire world.

Channel Pilot No. 27, Dover Strait Pilot No. 28, North Sea (West) No. 54, North Coast of Scotland Pilot No. 52, West Coast of England and Wales Pilot No. 37, and Irish Coast Pilot No. 40, cover the U.K. coasts.

Immediately inside each volume is a diagram showing the whole area covered by that particular volume, together with the limits and numbers of all the charts required for that specific area.

Each volume is normally republished at intervals of about 12 years, and between new editions it is kept corrected by publishing successive supplements every 18 months, each new supplement cancelling the previous edition. Small numbers of *Notices to Mariners* also are published each year to correct the 'Pilots', and in the chart correction notices the relevant page number of the 'Pilot' affect is quoted, so that a reference to this can be made in pencil in the margin until the issue of the next 'Pilot' supplement.

Admiralty List of Lights and Fog Signals

These volumes give details of lights, light-structures, and fog signals available to the mariner throughout the world. Light vessels, light floats, and light buoys, exhibiting lights at elevations exceeding 8 metres (26 feet) are included. Volume A covers the U.K. coasts. Volumes are republished at intervals of about eighteen months, but should be kept corrected in between new editions from Section 5 of the weekly editions of *Notices to Mariners*.

Admiralty List of Radio Signals

These give details of world-wide radio information and consist of six volumes as follows: Vol. 1 (*Communications*), Vol. 2 (*Navigational Aids*), Vol. 3 (*Meteorological Services*), Vol. 4 (*Meteorological Observation Stations*), Vol. 5 (*Miscellaneous Services*), *Admiralty List of Port Radio Stations and Pilot Vessels*. New editions of these volumes are published periodically, but they should be corrected from Section 6 of weekly editions of *Notices to Mariners*.

Admiralty Tide Tables

These are published in three volumes annually as follows: Vol. 1 *European waters* (*including Mediterranean Sea*), Vol. 2 *Atlantic and Indian Oceans*, Vol. 3 *Pacific Ocean and adjacent seas*. These tide tables are subject to constant and continual revision and information from obsolete editions should never be used.

Other very useful and informative publications are *Ocean Passages for the World*, *Admiralty Distance Tables*, and the *Nautical Almanac*.

Apart from the above Admiralty publications various almanacs, tide tables, *etc* are published by several firms in a form readily acceptable to fishermen, and one should use whatever publications are found most suitable for individual purposes.

Tidal Information

Further important information on a chart refers to *Tidal Information and Chart Datum* and to *Tidal Streams'*.

From *Chart 5063* will be seen such information as in Fig 25.

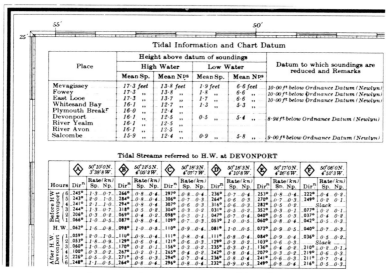

Fig 25   Tidal Information and Chart Datum

On examining a chart there will be seen in certain selected positions, diamond shapes with a letter inserted. These are called 'Tidal Diamonds'. At the position indicated the tide will be setting in the direction and at the rates referred to in the tables. By referring to the time of high water at Devonport (or any other port referred to in the tidal information) on that day one can find the direction in which the tide is setting, and at what rate at the position in which one may be situated relative to the nearest tidal diamond.

Set and Drift of Tides

It cannot be emphasised strongly enough that close attention should be paid to the set and drift of the tides. This will be dealt with in detail in the

DEPARTMENT OF TRADE
AND INDUSTRY

MERCHANT SHIPPING NOTICE No. M635

MEDICAL TRAINING OF DECK OFFICERS IN THE MERCHANT NAVY AND FISHING FLEET

Notice to Shipowners, Masters, Officers and Seamen of Merchant Ships and Fishing Vessels

This Notice supersedes Notice No. M.614

1. The medical training of deck officers in the Merchant Navy and Fishing Fleet is closely related to the *Ship Captain's Medical Guide*, in order to assist officers to acquire a good working knowledge of the Guide for use in dealing, when necessary, quickly and effectively with emergencies requiring medical treatment which arise at sea. The requirements which are described below provide for the issue of two grades of certificate, namely the First Aid at Sea certificate and the Ship Captain's Medical Training certificate.

BOARD OF TRADE

MERCHANT SHIPPING NOTICE No. M.587

SPECIAL SIGNALS FOR FISHING VESSELS

Notice to Owners, Masters, Officers and Seamen of Merchant Ships and Owners, Skippers, Mates and Crew of Fishing Vessels

The Maritime Safety Committee of the Inter-governmental Maritime Consultative Organization have considered and approved proposals for special signals for fishing vessels on a permissive basis as follows:

1. General

(a) Whilst the International Regulations for Preventing Collisions at Sea must be obeyed as appropriate, the signals herein are intended to prevent damage to fishing gear or accidents in the course of fishing operations.

(b) The light signals herein apply in all weathers from sunset to sunrise, when vessels are engaged in fishing as a fleet and during such times no other lights shall be exhibited, except the lights prescribed in the International Regulations for Preventing Collisions at Sea and such lights as cannot be mistaken for the prescribed lights or do not impair their visibility or distinctive character, or interfere with the keeping of a proper look-out. These lights may also be exhibited from sunrise to sunset in restricted visibility and in all other circumstances when it is deemed necessary.

Fig 26   Two specimens of 'M' Notices

38

chapters covering chart work. SET means the direction in which the tide is setting, and the DRIFT is found by multiplying the rate of the tide by the time. For example, if the rate of the tide is given as 2 knots over a period of 3 hours the drift would be 6 miles.

In the *Tidal Streams* tables it will be seen that the rate is given for springs and neaps, and in some cases there is quite a difference, so care must be taken to use the correct rate, or interpolate between them.

Merchant Shipping Notices

These are issued as required by the Department of Trade (previously the Board of Trade), and information and instructions to Owners, Skippers, Mates, and Crew of Fishing Vessels regarding such matters as safety, training, issue of certificates, *etc*. Each year a summary of notices still operative is issued, and it is necessary that one should be aware of such contents when they are applicable to fishing vessels. Not only may a question be put about them at the examinations but day to day work on board requires a good working knowledge of the various statutory requirements of the Department of Trade. Brief extracts of two 'M' notices (as they are known) are given in Fig 26.

Chapter 3   Chartwork

Chartwork is concerned primarily with coastal navigation, and having
learned about the use and construction of charts in the previous chapter
one should now be concerned with the practical application of chartwork.

Use of Instruments

First obtain practice in the use of instruments used in chartwork, which
include parallel rulers, dividers, a ruler, a pair of compasses, and a
protractor.

Parallel Rulers

There is a choice of types made in plastic or boxwood, amongst the best
known being *Captain Field's Improved.* These are 'walked' across the
chart from the compass rose to lay off a bearing or a course. It needs a
little practice to do this but one will become proficient as long as it is
remembered that when 'walking' them across a chart they should be
pressed firmly on the fixed portion with the fingers well spread, and move
the free part. Then press on this part and move the other part up to it.
For a test of proficiency lay your parallel rulers along a parallel of latitude
and 'walk' them to the next parallel of latitude above or below, and if it
lies exactly along it then the use of these rulers is being mastered.

As these parallel rulers are graduated also on the top and the sides as a
protractor it is often much more convenient to lay off true courses and
bearings by using them as a protractor, rather than working from a
compass rose which might be some distance away. To do this place the
index on the lower edge of the parallel rulers on to the nearest meridian
and rotate them until the required true course or bearing on the graduated
scale coincides with the meridian. This method requires practice but
ability to lay off courses and bearings is speedily acquired.

Made of brass, boxwood or plastic the roller type of ruler is preferred by some fishermen, but although easier to use they are apt to roll away or fall to the deck if let go when at sea in small craft. The *Douglas Protractor* is another useful device for laying off courses and bearings and is used in a similar manner to a protractor.

Dividers

These are used for measuring distances on charts and for plotting and taking off positions. Made of brass it is advisable to have a good pair of about 6 ins. long.

Compasses

These are useful pieces of navigational equipment made of brass into which a stub of pencil can be inserted.

Protractor

This instrument is made of plastic. It is a most useful item to include in navigational equipment as it can be used to lay off various angles, especially those involving horizontal sextant angles.

Plotting Positions

To plot a position place one edge of your parallel rulers along a convenient parallel of Latitude, or along the top or bottom of the chart. The parallel rulers must be exactly parallel to the horizontal line upon which it is laid, and the ruler is then moved until its upper edge reaches the required latitude on the vertical latitude scale on either side margin. Draw a line at the correct level on this scale and another line about 2 ins. long in the vicinity of the longitude required. A quick glance at the top or bottom margin of the chart will give the approximate longitude required. Now take the dividers and working from either the upper or lower margin of the chart measure the required longitude off from the nearest meridian. With this measurement on the dividers mark off the longitude on the line drawn already to mark the latitude. The point where the latitude is drawn and the point marking the longitude meet is the position required.

To Take Off a Position

This is a much simpler operation because dividers are used throughout (see Fig 27). First measure from the position to the nearest parallel of latitude, and then transfer the dividers to the latitude scale on the margin at either side of the chart and read the latitude. Next measure to the nearest meridian and read off the longitude from the scale at the top or bottom of the chart. When putting down Latitude and Longitude always put down the Latitude first.

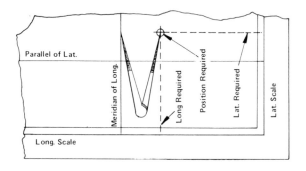

Fig 27 'Taking off' a position

Notations on The Chart

Dead reckoning, Estimated, and Observed positions.

The dead reckoning (DR) position is that obtained by plotting on the chart the course steered and the speed through the water. The symbol to make on a chart for the DR position is X.

The estimated position (EP) is that obtained after allowances have been made for current, tide, or leeway (if any). It should be marked on the chart thus △.

The observed position (Obs) is that position on the chart where it is known the boat to be because it has been fixed by bearings, soundings, sextant angles, etc. and it is marked ⊙.

In marking any of these positions on the chart remember to place the time by the side of the symbol, for example X 1120 or △ 2250 or ⊙ 0920.

Good habits in chartwork should be developed by always putting the correct notation and time alongside everything placed on a chart. Always note the time against every bearing or position plotted, and with any alterations of course. Against every course drawn on the chart its direction should be noted thus 035° (T) for a true course, 070° (M) for a magnetic course, and 298° (C) for a compass course.

A single bearing or position line should have a single arrow at the end thus

⟶

A transferred position line should be shown with a double arrow at each end thus

⇐⟷⇒

The direction of a tidal stream by three arrows on the line thus

Course to steer with one arrow on the line thus

Course to make good with two arrows thus ————➤➤————

Courses and Distances

To use parallel rules for laying off a course (Fig 31, page 46) place the bevelled edge of the ruler on each of the two positions and join them with a pencilled line. If the course has to be laid to pass a certain distance off a lighthouse or point of land, *etc.* first draw a circle with the compasses, using the distance off as a radius, using the lighthouse, *etc.* as the centre. Then draw the course line so that it just touches the outside edge of the circle, or as a tangent to it (see Fig 28).

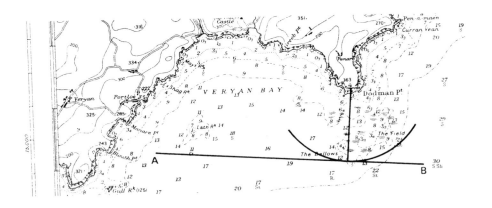

Fig 28   Marking course and distance

Once the course line is drawn move the parallel rulers to the centre of the nearest compass rose and read off the true course from the outer ring. Remember that the parallel rulers can be used as a protractor and read off the true course from the nearest meridian as described earlier in this chapter.

To measure the distance use the dividers and by measuring the length of the course line transfer the dividers to the latitude scale on either side of the chart, remembering to use this scale in the same latitude as one is working.

Note that if the distance to be measured is greater than the maximum spread of the dividers put a convenient measurement on them from the latitude scale such as 10, 15 or 20 miles, and measure off these units along the course line. The final short leg can be measured separately and added

43

to the distance already measured. Study the following chart examples (Figs 29 and 30) of a course and distance between two points.

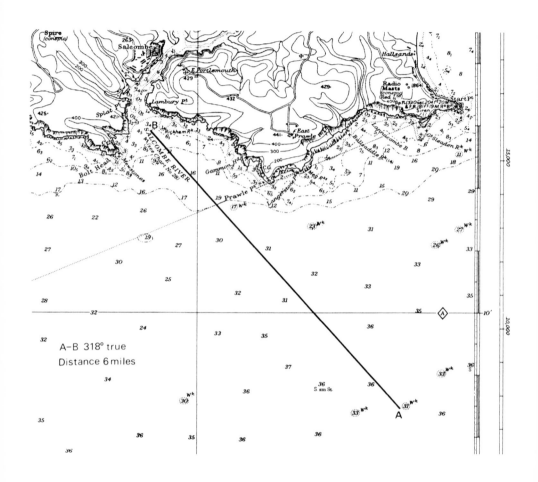

Fig 29   Example of marking of course and distance

Bearings and Cross Bearings

One can fix a position on a chart by various means, and in order to maintain the course laid down on the chart a check should be made on position frequently by observations of terrestrial objects.

A 'position line' is any line drawn on a chart on which one's position is known to be, but a 'fix' is the interception of two or more position lines which have been obtained at approximately the same time.

44

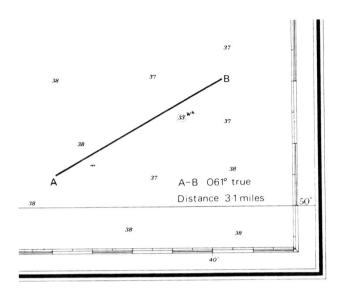

Fig 30   Further example of marking course and distance

The methods of obtaining a position line can be summarised as follows and is demonstrated in Fig 32 (see following pages for Figs 31 and 32).

1   The visual bearing of a terrestrial object.
2   A transit bearing (that is two objects in a line).
3   The radio bearing of a radio beacon.
4   A circle of position obtained from the vertical sextant angle or from horizontal sextant angles.
5   Horizontal angles obtained by taking compass bearings, whether the compass error is known or not.
6   A circle of position obtained by the range of a light seen to rise above, or dip below the horizon.
7   A sounding, or a line of soundings.

To obtain a fix one must:

1   Obtain two or more visual bearings (that is cross bearings preferably with 90° angle between them).
2   A transit bearing and the visual bearing of another object.
3   Two or more radio bearings.
4   Horizontal angles between three terrestrial objects (either by the use of a sextant or the use of compass bearings).

45

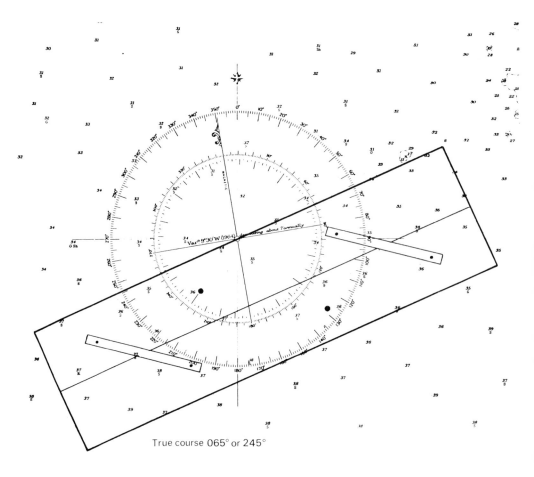

True course 065° or 245°

Fig 31 Use of parallel rulers in laying a course

5 The bearing of a light, just dipping, together with the maximum range of the light at eye level.
6 The visual bearing of a terrestrial object and the distance off by a vertical sextant angle.
7 A sounding and the bearing of an object.

For demonstration see Fig 33, page 48.

Running Fixes

When only one suitable landmark or light is visible a bearing of this will give a position line. Take this first bearing and lay it off on the chart. The

46

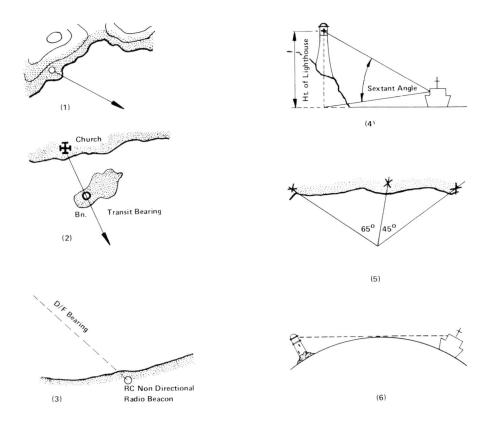

Fig 32 Checking by bearings and cross bearings

boat must be somewhere on this bearing. Proceeding on the course another bearing is taken of the same landmark or light (or another object if the first object is lost to sight). By this means a 'running fix' is obtained which simply means transferring the first position line (or bearing) to cut the second position line or bearing.

An accurate fix can be obtained by this means provided it is known within reasonable limits, the estimated track and distance made good between the two bearings (see Fig 34). It is advisable to try and make the distance run between the bearings as short as possible provided the angle of the cut is as near as possible to 90°.

The order of work is as follows:

1 Lay off the two bearings obtained A P and B P.
2 Mark where the first bearing cuts the course line (one must be somewhere on this line which is a single position line, and call it K). (*Continued on page 49*)

(*Continued on page 49*)

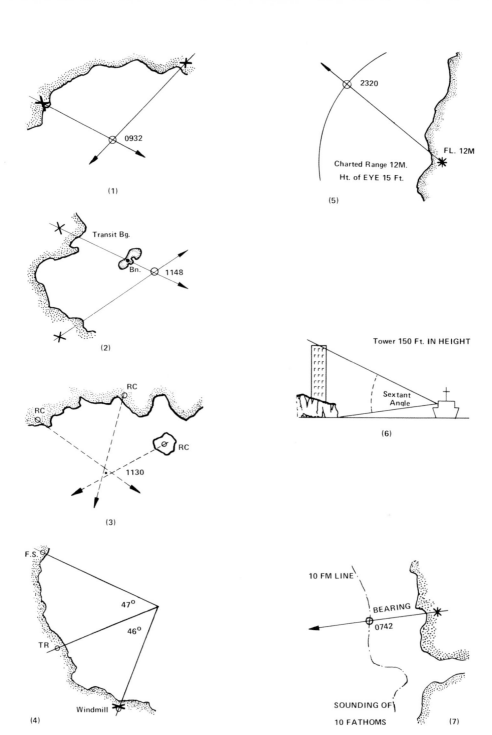

Fig 33   To obtain a 'fix'

48

3 From K lay off along the course line the distance steamed K Y.
4 Through Y lay off the first bearing as a transferred position line and where it cuts BP at O is the boat's position.

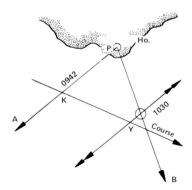

Fig 34 Securing a 'running fix'

If, between the two bearings it is known that the boat's course and speed through the water is being affected by a tide or current one must lay off the tidal stream from position Y, and draw a transferred position line through the end of the tidal stream.

The order of work at the commencement would follow that of above, but after laying off the course and distance steamed K Y one must lay off the tidal stream from Y in the direction and at the rate it is known to be flowing Y T. When transferring the first bearing A P it must pass through the point T and where it cuts the second bearing B P at O is the boat's position as shown in Fig 35.

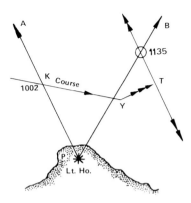

Fig 35 Allowing for tidal stream in fixing position

Setting a Course to Counteract the Effects of a Tide

To navigate correctly and efficiently one should not allow the boat to be set off the course line laid on the chart. Provided the direction and rate of a tidal stream, or tidal streams, is known at the commencement of the passage (and throughout the passage) a course, or courses, must be set to counteract such tidal streams (see Fig 36).

By the simple method of counteracting the tide the boat should remain more or less on the course line throughout the passage, but circumstances do arise when the strength and direction of these tidal streams may not be exactly as predicted, in which case it may be found that the boat is slightly 'off course,' but as tides are only *predicted* and are not exact this is a risk which must be foreseen. By checking regularly the position of the boat when on a coastal passage it can always be seen whether the correct allowance for the tidal streams have been made and course adjustments made as necessary. Order of work. See diagram below.

1 Lay off the course it is required to make good A B and mark with two arrows 275°.
2 From A lay off the direction of the tidal stream 025° A T and mark it with three arrows. Along this line mark the tide for a period of one hour, or two hours, whichever is convenient (in the example one hour is used at an appropriate scale) and mark the position T.
3 From T using compass or dividers set a radius equal to the distance the boat would steam within the period of time used, that is one hour, or two hours, *etc* as long as it is for the period of the tide. Draw a small arc where this cuts the course line to be made good A B, and make a mark C.
4 Draw a line from T to C, marking it with one arrow, and the direction of this line T C is the true course required to steer from A to counteract the effect of the tidal stream setting in the direction A T.
5 From A to C is the actual distance the boat will make good within the period of time.

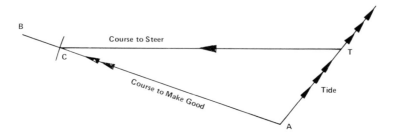

Fig 36   Setting a course to counteract tidal stream

To Find the Set and Drift of the Tide

This is a simple problem which involves 'dead reckoning' position by plotting the course and distance steamed from a given position in a certain time. Then cross bearings are given to find the 'observed position.' When this has been plotted draw a line from the DR position to the Obs position, and this line will give the direction (or set) in which the tide has flowed, and the drift will be the distance between the two positions. If asked also to find the rate of the tide divide the amount of the drift by the time interval that has elapsed between leaving the first departure position and the time of the observed position (Fig 37 shows the process). A drift of 4 miles in 2 hours would be $\frac{4}{2}$ which equals 2 knots (2 nautical miles per hour).

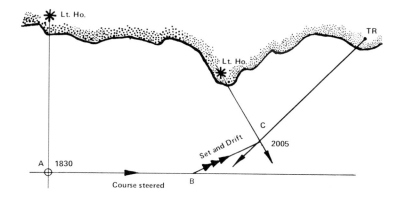

Fig 37   Finding set and drift of the tide experienced

A   Departure position
B   DR position at the end of the period steamed.
C   Observed position.
BC True direction (set) of the tide and the drift is the distance in nautical miles measured from the latitude scale.

Speed, Time and Distance

In chartwork ability to work out speed is essential, as well as the distance covered, and the time it takes to cover a given distance. The formula to find these three factors is quite simple.

To find *speed* divide the *distance* made good by the *time* taken, *eg* if 12 miles is travelled in 2 hours the speed would be 12 divided by 2, which equals 6 knots.

To find the *time* taken to cover a given distance divide the *speed* into the *distance*, *eg* if it is required to know how long it would take to cover 15 miles at 5 knots divide 15 by 5 which would equal 3 hours.

3

Distance covered is found by multiplying *speed* by the *time* taken, *eg* if a boat steamed for 4 hours at 7 knots it would mean 4 multiplied by 7 equals 28 miles covered.

Summarised it would look like this:

$$Speed \quad equals \quad \frac{Distance}{Time}$$

$$Time \quad ,, \quad \frac{Distance}{Speed}$$

$$Distance \quad ,, \quad Speed \times time$$

In the above examples round figures are used, but in practice at sea all or any of the three factors in parts of a knot, hour, or nautical mile may be involved. Decimals are required in solving problems that do not involve a round figure, *eg* if a boat had a speed of 5·5 knots and sailed for 2½ hours it would be 5·5 × 2·5 equals 13·75 miles.

In changing time into decimals work to the nearest 6 minutes which is one-tenth or ·1 of an hour. In slow speed boats it is quite accurate enough to work to one place of decimal, but for examination purposes it is advisable to work to two places of decimals. For example if a boat sailed 22 miles in 4 hours 12 minutes and one wished to know its speed it would be 5·24 knots to the nearest decimal.

Try some worked examples:

What is the *Speed*

1. Distance 18 miles in 3 hours 36 mins. *Answer* 5 knots
2. Distance 22 miles in 2 hours 48 mins. *Answer* 7·9 knots

What *Time* would it take to travel the following distances

3. Distance 31 miles at 5 knots *Answer* 6·2 hours or 6 hours 12 minutes
4. Distance 19 miles at 6·5 knots *Answer* 2·92 hours or 2 hours 55 minutes

What *Distance* would be covered at the following speeds and time

5. Speed 4·5 knots in 4 hours 48 minutes *Answer* 21·6 miles
6. Speed 6·7 knots in 3 hours 18 minutes *Answer* 22·11 miles

Leeway

A fishing boat can be pushed sideways by the pressure of the wind upon the hull and superstructure. Such pressure can be exerted when the wind is in any direction, relative to a fishing boat except when it is directly ahead or astern.

This sideways movement is known as leeway, and is expressed as an angle in degrees, and is the angle between the course the boat is steering and the actual track being made good through the water. Leeway depends so much upon the direction and strength of the wind, the draft and trim of the boat and the amount of freeboard and superstructure, that no specific set of rules can be laid down as to how much leeway would be made in any given set of circumstances.

Only by observation, and that only approximately, can one determine the amount of leeway a boat is making (see Fig 38). By observing the farthest visible part of the wake (and a towing log line if carried) and comparing this with the fore and aft line of the boat, one can estimate the degree of leeway. Another method is to take a compass bearing of the wake and reverse it to estimate the course being made good through the water.

For example, steering 090° by compass, wind north, estimated leeway 10°, estimated compass course being made good 100°.

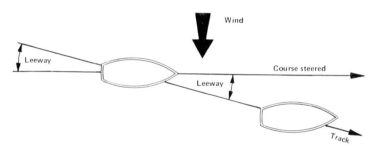

Fig 38   Checking leeway

Wake would have had a compass bearing of 280° which reversed makes 100°, or by allowing 10° *away* from the wind 90° plus 10° equals 100°.

A good navigator, once having estimated leeway would normally counteract it by steering a course, equivalent to the amount of leeway, to windward. In the above example, a compass course of 080° would counteract the wind to the extent of the 10° leeway.

It must be remembered however that when counteracting for leeway always apply this to the true course and not to the compass course, because an alteration for leeway made to the compass course could alter the deviation, and the compass heading might be changed.

For example. with a true course of 270° wind north and 10° leeway it would be necessary to steer a course of 280° and with variation of 8° west would make a magnetic course of 288°. To obtain the correct deviation for this magnetic course refer to the deviation card. If the variation had been applied to 270° it would make a magnetic course of 278° which would produce a different deviation, so always remember to apply leeway to the true course, followed by the variation and the deviation.

Chapter 4   The Sextant

Description of the Sextant — Principles of Construction — Index Error — Adjustments — Measurement and Correction of Altitudes — Horizontal and Vertical Sextant Angles.

Description of the Sextant

The following is a technical description of the sextant (shown in Fig 39).

a   The index bar which is free to rotate on a central axis under the mirror D.

b   Micrometer vernier for quick and accurate reading.

c   Graduated arc of the sextant.

d   Is a mirror called the index glass.

e   Is a mirror called the horizon glass.

f   Telescope.

g   Tangent screw for making the finer readings on the micrometer vernier.

h   Coloured shades for index and horizon mirrors.

i   Collar for holding telescope.

j   Small screw for adjusting index glass to make it exactly perpendicular to the plane (surface) of the sextant.

k   Screw (small) to adjust the horizon glass to make it exactly perpendicular to the plane of the sextant.

l   Small screw to adjust the horizon glass to make it exactly parallel to the index glass when the vernier is set at zero on the scale.

m   Screw to adjust height of telescope collar.

n   Index bar clamp which holds the bar into position when it is released by locking it into the grooved rack behind the arc.

Principles of Construction

The sextant is used for measuring the angle, at the observer's eye between:

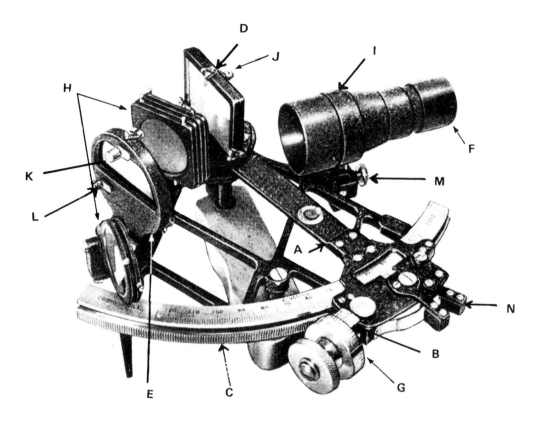

Fig 39   The sextant

1   The top of some object such as a lighthouse, tower, *etc.*, and the water line when the sextant is held vertically.
2   Two objects on different bearings on shore, with the sextant held horizontally.
3   A celestial object (sun, moon, stars, or planets) and the horizon.

The sextant is an instrument of double reflection, the angle measured by it being double the angle between the two reflecting mirrors. As the index glass is fixed to the index bar, the angle measured is double the distance which the index is moved along the graduated arc from 0°. It will be seen that the sextant is normally graduated on the arc to read 120°, although the actual angle between the mirrors does not exceed 60°.

Fig 40 shows how the double reflection and the measurement of the angle take place, although a knowledge of the optical principles involved is not normally required for examination purposes.

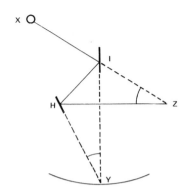

Fig 40   Diagram showing double reflection of sextant to secure measurement of angle

X is the sun
I is the index glass (mirror)
H is the horizon glass (mirror)
Z is the observer's eye
Y is the angle measured on the arc of the sextant

The altitude of the sun is angle XZH, which is twice the angle IYH on the arc of the sextant. Because it is an instrument of double reflection, the angle measured by it is double the angle between the reflecting mirrors. As the index glass is fixed on the index bar, the angle is double the distance which the index is moved along the arc from 0°. When the index is set at 0° the index glass and horizon glass are parallel to each other.

Index Error

There are a number of instrumental errors which can occur in a sextant but of these only index error is required to be known by candidates for the Second Hand Certificates, although a brief mention of two other important errors is made.

The first is called 'error of perpendicularity' and is caused by the index glass not being perpendicular to the plane (surface) of the instrument.

The second is called 'side error' and would be caused by the horizon glass not being perpendicular to the plane (surface) of the instrument.

The third is known as 'index error' and is caused by the index glass not being parallel to the horizon glass when the index is at zero, and is the error with which an examinee is concerned.

From these three errors it will be seen that if a sextant is to measure angles accurately the index and horizon glasses (mirrors) must be perpendicular to the plane (surface) of the sextant, and when the index is at zero the two glasses (or mirrors) must be parallel to each other.

56

To find and correct these three errors use is made of what are called the first, second and third adjustments respectively. It is the 'third' adjustment—to find and correct index error—which is now explained.

To make the 'third' adjustment clamp the index of the sextant at zero and holding the sextant vertically look through the telescope and horizon glass at the horizon. Should the true and reflected horizons appear in one continuous line the horizon glass is parallel to the index glass and no 'index' error exists.

If the true horizon appears above or below the reflected one the horizon glass is not parallel to the index glass.

Study the drawings below and you will see how "index error" is shown.

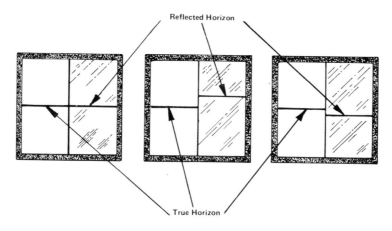

Fig 41   Illustration of 'index error'

To make the necessary adjustment to correct 'index error' turn the small screw gently at the back of the horizon glass (the screw nearest to the plane of the instrument 'l' in the illustration of the sextant shown at the beginning of this chapter) until the true and reflected horizons are in one line.

If the sextant is not fitted with a screw for making this adjustment, hold the sextant vertically, as described previously, and move the tangent screw ('g' in the illustration) until the true and reflected horizon appears as one straight line. The reading will be the amount of 'index error' which must be added if the reading was off the arc and subtracted if on the arc.

For example, if the reading was 2′ 30″ off the arc add that amount to all future angles taken with the sextant, because it means that every angle measured is reading that much too low. If it was 3′ 30″ on the arc subtract that amount from any future angles measured.

The sextant should be checked frequently for 'index error'.

Should a clear horizon be unobtainable to check for 'index error' two other methods can be used. First, at night sighting on a star clamp the index at zero, hold the sextant vertically and look through the telescope and horizon glass at the star. If the star and its reflection exactly coincide no 'index error' exists, but if one star appears above the other (or below it) then there is an 'index error'. Adjustment can be made as described above until the true and reflected star coincides and when only one star can be seen, or, sighting on the horizon, move the tangent screw until the same effect is obtained, reading the amount of 'index error' on or off the arc.

| No index error | * True | * Reflected |
| * | * Reflected | * True |

A straight horizontal line must never be used such as the roof of a house or shed, *etc*, instead of the sea horizon, because it would introduce a factor called 'parallax', which is caused by the index glass and the horizon glass not being on exactly the same plane when the sextant is held vertically to look at the horizon. It has no visible effect when the sea horizon is used because it is so far distant.

If the horizon or a star is not visible the 'index error' can be found by using the sun as shown in Fig 42. Clamp the index about 32' off the arc, hold the sextant vertically and look through the telescope and horizon glass at the sun. Two suns will appear, one above the other, one of which is the true sun and the other the reflected image. Turn the tangent screw until the limbs of the two suns are in exact contact and note the reading. Next clamp the index about 32' on the arc and carry out the same procedure.

With the two readings subtract the lesser from the greater and divide the difference by two. The result will be the 'index error', to be subtracted if the greater reading is on the arc, but added if the greater reading is off the arc.

An example of this would be as follows:

Sextant reading 35' 30" off the arc
    ,,   ,,   30' 30" on the arc
_____

Divided by 2   5' 00" difference
    =   2' 30" equals index error to be added because the greater reading is off the arc.

The sextant is set to about 32' on and off the arc because that is the approximate diameter of the sun as seen from the earth.

It is not advisable to try and find 'index error' by using the sun when the

58

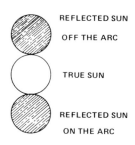

REFLECTED SUN
OFF THE ARC

TRUE SUN

REFLECTED SUN
ON THE ARC

Fig 42   Using the sun to find 'index error'

altitude is very small because of excessive refraction, so try and await a favourable time when the sun has reached a moderate altitude, say about 20°.

Measurement and Correction of Altitudes

At this stage candidates are advised to obtain the use of a sextant, to study its details in conjunction with the contents of this chapter, and to learn to handle and read it.

When picking up a sextant, either from its box or from the wheelhouse or chartroom table, always graps it first with the left hand by putting three or four fingers firmly into the framework of the plane of the sextant, lifting it up and transferring it to the right hand which should grasp the handle which is situated beneath the plane of the instrument, or on the right hand side when it is held vertically.

When taking a 'sight' make sure the correct shades cover the index and horizon mirrors. Darker ones if the sun and horizon are brightest, and lesser shades if the sun and the horizon are fainter. To bring the sun down to the horizon set the index at zero and look directly at the sun, moving the index bar slowly away so as to bring the sun downwards until the lower limb touches the horizon. If the approximate altitude of the sun is known, this can be set on the arc of the sextant and then look at the horizon directly beneath the sun. This, when observed, in the horizon glass can be adjusted by means of the tangent screw until its lower limb touches the horizon. To make certain of an accurate reading of the sun's lower limb on the horizon move the sextant slightly with a pendulum motion which will cause the sun to move from side to side like a pendulum, and at its lowest point it should just touch the horizon.

To read the sextant, first note the index mark which indicates the number of degrees on the graduated arc ('c' in the illustration) and the arrow on the vernier scale indicates the number of minutes on the revolving vernier, while the seconds (to the nearest 10) are read on the smaller vernier scale at the side.

59

Some specimen micrometer vernier readings are shown in Fig. 43.

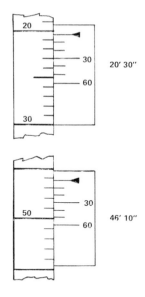

Fig 43   Specimen micrometer vernier readings

When taking the altitude of the sun (or a 'sight' as it is called) by using the sextant the angle observed is between the lower limb of the sun (or in exceptional cases the upper limb if the lower limb is obscured by cloud) and the horizon, that is the visible horizon, the one seen from the bridge of the vessel (see Fig 44).

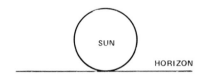

Fig 44   Sighting the sun

Before the angle thus obtained can be used it must be corrected for various factors because what is required finally is the true altitude, which is the true angular height of the sun's centre at the centre of the earth.

60

Study Fig 45 and learn the definitions as listed here:

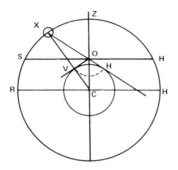

Fig 45   Finding true altitude

C	the centre of the earth
O	an observer
Z	the zenith
X	the sun
VH	visible horizon
SOH	sensible horizon
RCH	rational horizon
SOV	dip
XOV	observed altitude
XOS	apparent altitude
XCR	true altitude
OXC	parallax

The Visible Horizon

This is the circle bounding the observer's view at sea. VH is the horizon as seen by the observer at O.

The Sensible Horizon

SOH is a circle whose plane passes horizontally through the observer at O, and is at right angles to the vertical. COZ is the vertical and SOH the plane of the sensible horizon.

The Rational Horizon

RCH is the circle whose plane passes through the centre of the earth parallel with the sensible horizon. RCH is the plane of the rational horizon.

Dip

This is the angular depression of the visible horizon below the sensible horizon and the sketch (Fig 46) shows how dip is defined. It will be seen from this that dip (Fig 46) depends upon the height of the eye above sea level, and the greater the height the larger the angle of dip.

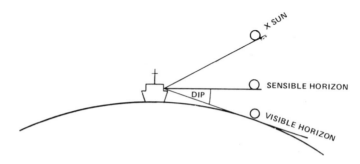

Fig 46   How 'Dip' is defined

The sea, or visible horizon, is the only one that can be seen, and the sensible and rational horizons are imaginary ones introduced for astronomical convenience to assist in the correction of altitudes, the centre of the earth being the common point of reference to which all observed altitudes must be reduced before the working of the sight can be used.

Sextant Altitude

This is the angular height of an object above the visible horizon as read from the arc of a sextant before correction for index error (if any).

Observed Altitude

This is the angular height of an object above the visible horizon as read from the arc of a sextant and corrected for index error.

Apparent Altitude

This is the angular height of an object above the sensible horizon after correcting the observed altitude for dip.

True Altitude

This is the angular height of an object's centre above the rational horizon at the centre of the earth.

Zenith Distance

This is the angular distance of a heavenly body from the zenith of the

62

observer, and is found by subtracting the true altitude from 90° (90° minus the true altitude equals zenith distance ZD)

Parallax in Altitude

This is due to the fact that the heavenly body is observed from the surface of the earth whilst it is necessary to find its angle from the centre of the earth. It is the angle at the object between a line drawn from the centre of the earth and one drawn from the observer. The nearer an object is to the earth the greater the parallax. The sun's parallax is very small and never exceeds 9″ of arc.

Semi-diameter

This is half the angular diameter of the sun as viewed from the earth. As it is necessary to find the angular height (altitude) of the centre of the sun the semi-diameter SD must be applied as a correction. If the lower limb is observed then the SD must be added but if the upper limb then subtract the SD. The sun's semi-diameter is obtained daily from the *Nautical Almanac*, or any other nautical publications that may be in use.

Refraction

This is the bending of the sun's rays as they enter the atmosphere of the earth and make the sun appear higher than it really is. Refraction is at its maximum when the sun first rises, and nil when it is overhead (see Fig 47, next page).

In practice the whole of the corrections to apply to an observed altitude are contained in the *Sun's Total Correction Table*, the factors required being the observed altitude, height of eye (given in either feet or metres above sea level) and the month of the year. The reason for including the month is that during the winter months the sun is closer than during the summer months, and the SD is therefore slightly larger during the winter.

In the *Nautical Almanac* the correction for dip is given in a separate table, and the total of the remaining corrections is given in an adjoining table, that is, parallax, refraction and semi-diameter. A careful check must be made of the month of the year in which the 'sight' is taken, and whether the lower or upper limb of the sun is observed.

Horizontal and Vertical Sextant Angles

In addition to measuring the angular height of heavenly bodies the sextant can be used to measure horizontal and vertical angles of terrestrial objects, and very quick and accurate results can be obtained by the use of such angles.

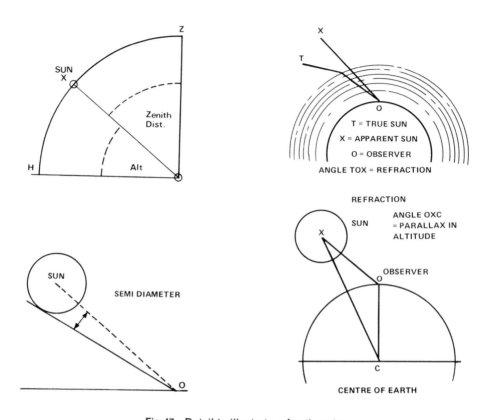

Fig 47   Detail to illustrate refraction etc.

To take a horizontal angle between two shore objects (or floating objects), hold the sextant horizontally and always work from right to left, that is observe the right hand object first, and move it sideways to the left towards the second object (see Fig 48). When holding it horizontally the right hand holding the handle is underneath the instrument.

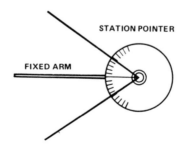

Fig 48   Diagram of a station pointer

If three identifiable shore objects are selected which are nearly in a line, then by measuring the horizontal angles between them, a very accurate fix of one's position can be obtained. Use either a station pointer, tracing paper, or by laying off the angles geometrically by the use of a protractor to fix a position.

A station pointer is made of brass or plastic and has three arms attached to a circular graduated plate. The centre arm is fixed and the two other arms rotate so that the angles between the centre objects and the objects to the right and left can be set on the graduated plate and clamped into position. The station pointer is moved around on the chart until the bevelled edge of each arm coincides with each selected object (there is only one position in which this can occur). When the station pointer is so placed make a mark on the chart through a hole in the centre of the graduated plate, and that is the position.

By using tracing paper it is possible to plot similar angles and adjust the tracing paper over the three selected objects until they all coincide, and by making a mark on the chart through the tracing paper it will indicate position (see Fig 49).

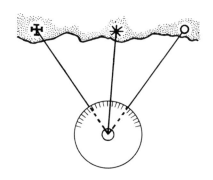

Fig 49 Using a station pointer to obtain a fix

In using a protractor instead of a station pointer join the three selected positions by straight lines, and lay off from each position an angle which is equal to the *complement* of the horizontal angle between the two positions. The complement is obtained by subtracting the angle from 90°. Where these two lines intercept is the *centre* of a circle passing through both points of land and the ship. As the ship is on *both* these circles it must be at their point of interception as shown in Fig 50.

If lines are drawn from this point of interception to the points of land they will be found to contain the angle measured (horizontally) with the

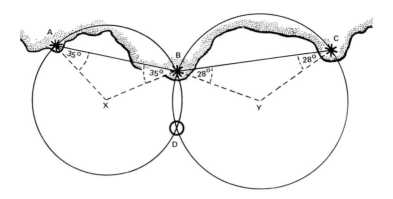

Fig 50   Using a protractor to obtain a fix (angles not drawn to scale)

sextant between the points. From the above drawing the horizontal angle measured with the sextant between A and B was 55°, therefore the angle to lay off from these points is 90°–55° equals 35°. Between B and C the angle was 62° therefore the angles would be 90° minus 62° equals 28°. X and Y are the centres of the circles to be drawn and as D is the point where they meet (intersect) that is the boat's position. If the angles at D are drawn between A and B, and B and C, they would be the horizontal angles measured by the sextant.

It will be noted that the angles from the points on land have been drawn on the *seaward* side, but if the angles between them had been over 90° it would be necessary to subtract this from the horizontal sextant angle and draw the resultant angles on the *opposite* side of the line. For example, if the angle between the points had been 110° the angle drawn would be 110° minus 90° equals 20°.

The best results are obtained when the three chosen objects are in or near the same straight line, or the boat is inside a triangle formed by the three objects. It is also advisable that the centre object be nearer than the other two.

To take a vertical sextant angle of a shore object hold the sextant vertically, and looking at the top of the object move the index bar slowly as it is brought down to the water-line immediately below (or by using the tangent screw if the angle is very small). Always be careful when measuring the vertical height of a lighthouse that the centre of the lens of the light is taken (which would normally be clearly visible through the sextant telescope). It is always the height of the light that is given above sea level and not the top of the lighthouse. The height of the object above sea level must be known before the distance off can be found.

In a right angled triangle given the length of one side and one of the lesser angles, one can calculate the length of the other side, but tables are

66

available in various nautical publications giving an immediate answer. Usually they are headed 'Distance by Vertical Angle' or 'Distance by Vertical Sextant Angle'. Take from the tables with the height of the object in feet or metres and the angle obtained with the sextant, corrected for index error if any (see Fig 51).

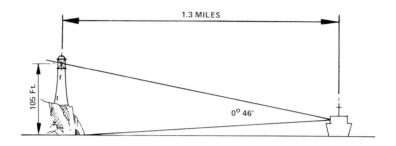

Fig 51   Securing distance by vertical angle

As an example, the sextant angle of an object 105 feet high was found to be 0° 46′. By looking at the table under 105 feet for height and 0° 46′ as the angle it will be seen that the distance off was 1 mile 3 cables, or 1.3 miles. In a further example when an object 80 feet high has a vertical angle of 1° 5′, the tables show a distance off of 7 cables, or 0.7 miles.

The height of shore objects such as lighthouses, *etc.* are given above mean high water springs, so that from the vertical angle at low water the object may be as much as 30 or 40 feet higher above the water-line than indicated by the height given on the chart. For accurate fixing add the amount the tide has fallen below mean high water springs to obtain the correct height of the object above the water-line, but if this is ignored and the charted height used, then a distance off will be obtained *nearer* to the shore object than it is in fact—which is, of course, a safety factor.

For example, if the vertical sextant angle of an object measured is charted as 115 feet high and the angle obtained was 1° 21′ the distance off by the tables would be 8 cables, or 0.8 miles, but if it was low water and the object was in fact 145 feet above the water line the correct distance off would be 1 mile, so one would be plotting one's vessel's position 0.2 miles nearer the object than it really was.

In problems involving vertical sextant angles an examinee might be asked how to proceed if required to pass, or round, a lighthouse, headland, or known object at a defined distance. This applies particularly if a reef or ledge of rocks runs out from a headland. It is essential that it is passed at a safe distance.

The procedure is to turn to the 'Distance by Vertical Angle Tables' and, with the height of the object at the top of the table, look down the left hand column until the figure giving the distance off it is wished to pass is found. Where these two factors meet in the columns giving the angles is the angle to set on the sextant.

As the object is approached, observe it through the sextant telescope and as soon as it is level with the water line immediately below the object one will have arrived at the distance off, pre-determined from the tables. Remember that where there is a very large rise and fall of the tide this should be allowed for in calculating the true height of the object above the water line.

As long as the object remains at this water line level the correct distance is being maintained, but if it falls below the water line then one is moving farther off. If it rises above the water line level one is moving closer to the shore. Study Fig 52.

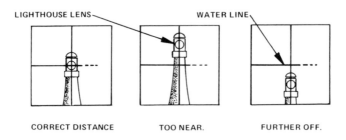

Fig 52 Checking correct distance by sextant

Chapter 5 Use of Traverse Tables
Plane and Mercator Sailing

DIFFERENCE OF LATITUDE — DIFFERENCE OF LONGITUDE — DEPARTURE — FINDING POSITION BY THE USE OF *Traverse Tables* — PLANE SAILING — PARALLEL SAILING — MERCATOR SAILING — MERIDIONAL PARTS.

It is essential to become thoroughly familiar with the definitions and the diagrams that follow (see Fig 53).

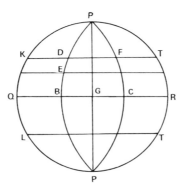

Fig 53 Definitions diagram (see text for detail)

Definitions

P P	The poles
Q R	The equator
P G P	The prime meridian
P B P	A meridian
P C P	,,
K T	A parallel of latitude
L T	,, ,,
G B	Longitude of P B P (Long).
G C	,, P C P
B C	Difference of longitude (D. Long.)
D F	Departure (Dep.)
B E	Latitude of E (Lat.)
C F	,, F
E D	Difference of latitude (D. Lat.)

69

Difference of Latitude (D. Lat.)

This is the number of degrees and/or minutes of latitude between two positions or places situated in different latitudes. For example, the D. Lat. *from* 49° 50′ N *to* 48° 45′ N is 1° 05′ which equals 65′ (1° equals 60′) S. This is equal to 65 nautical miles because one minute of latitude equals one nautical mile. From 48° 00′ N to 51° 00′ N the D. Lat. would be 180′ N.

When finding the D. Lat. between two positions be very careful to name it correctly. It is named according to the direction from one place to another, north if more northerly, and south if more southerly. Note how the two above examples have been named.

Difference of Longitude (D. Long.)

This is the number of degrees and/or minutes of longitude between two points or positions situated in different longitudes. The D. Long. expressed in arc remains the same in whatever latitude one may be. For example, from 10° W to 11° 45′ W is 105′ W (1° equals 60′). From 9° W to 7° W equals 120′ E (1° equals 60′). These figures would be expressed in exactly the same manner in whatever latitude one may be situated.

D. Long. is named according to the direction from one position to another. East if more easterly and west if more westerly. Care must be taken when the change is from a westerly longitude to an easterly longitude, and vice versa. For example, from longitude 2° W to longitude 2° E the D. Long. would be 240′ E, whereas from Long. 2° E to Long. 2° W it would be named 240′ W. It will be noted that the longitudes are added in these cases.

Departure (Dep.)

This is the east–west distance between two meridians on the surface of the earth measured along a parallel of latitude and expressed in nautical miles. On the equator only, departure (which is always expressed in nautical miles) is equal to the difference of longitude (D. Long.) expressed in minutes of arc, but immediately one leaves the equator and moves into higher latitudes (that is travelling north in the northern hemisphere, and south in the southern hemisphere) the meridians of longitude begin to converge until they meet at the poles.

The actual distance, expressed in nautical miles, begins to get less, and this distance is called departure (Dep.). It can be defined as the distance made good in nautical miles east or west of the point from the point sailed.

The distinction between D. Long. and Dep. must be clearly understood, and the conversion of one factor into the other is an essential part of the use of the *Traverse Tables*, and of the solution of the problems to follow.

Study Fig 54 and the explanatory text which together give a clear idea of the relationship between the three factors involved.

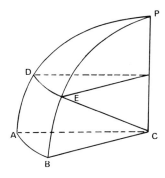

Fig 54 Diagram of factors involved in 'Departure'

P is the pole of the earth.
C is the centre of the earth.
P A and P B are two meridians.
Arc A B is the D. Long. at the equator.
Arc D E is the arc of a parallel of latitude between the two meridians, and is also the departure or actual distance between them in that latitude.

Using the Traverse Table

Instead of plotting courses and distances on a chart one can obtain D R positions by the use of the *Traverse Tables*, given the compass courses steered, variation and deviation (thence the true course) and the distance run recorded by log or calculated by estimated speed. Also, it is possible to calculate the allowance for tides and leeway, if any.

The departure position must be known either as a given Lat. and Long. or by the bearing and distance from a point of land. In addition, by using the tables one can calculate the course and distance between two points by plane sailing.

These tables give a quick solution to right-angled plane triangles for every degree from 0°–90°, and for units of distance up to 600 miles. A few simple diagrams (Fig 55 and others) explains how the *Traverse Tables* are used to make the working simple and easy.

How Traverse Tables are used

It will be seen clearly from the drawings B and C that the greater the latitude the smaller the departure relative to the D. Long.

71

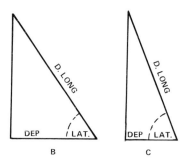

Fig 55   Diagram showing how traverse tables are used

Turn to the extract from the *Traverse Tables* at the back of this book and it will be seen that the various headings given in these tables consist of the terms just explained and shown in the drawings.

Dist., D. Lat. Dep. and D. Long. The degrees at the top and bottom of the page, 28° and 62°, respectively show either the course or the mean latitude according to what particular problem one is solving.

If a true course of 028° is steered, distance 72 miles, using the tables from the top of the page, it will be seen that D. Lat. would be 63.6′, and the Dep. 33.8′. If the true course was 062° distance 46 miles, working from the bottom of the page, the D. Lat. would be 21.6′ and Dep. 40.6′. (Fig. 56.)

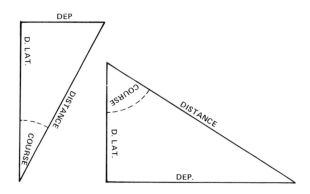

Fig 56   Further diagram showing how traverse tables are used

Although the tables are tabulated from 0° to 90° it will be observed that at the top and bottom of each page the three figure notation is given on each side of arrows pointing upwards. Easterly courses are on the right

72

hand side of the arrow, and westerly courses on the left. For example, using 28° can indicate a true course of either 028° or 152°, or a course of either 332° or 208° according to which direction one is heading.

Some problems could give the course in quadrantal graduations, or the three figure notation, but whatever method one is called upon to use be very careful to get the courses named correctly. For example, if the course is 282° look in the *Traverse Tables* above that figure, or N 78° W, or for 189° look under S 9° W or 189°.

Give the D. Lat. and Dep. for the following courses and distances.

						Answers		
Course	028°	Dist.	50 M	D. Lat.	44.1′ N	Dep.	23.5′ E
,,	152°	,,	80 M	,,	70.6′ S	,,	37.6′ E
,,	118°	,,	160 M	,,	75.1′ S	,,	141.3′ E
,,	298°	,,	60 M	,,	28.2′ N	,,	53.0′ W

To complete the use of the *Traverse Tables* one must learn how to find the difference of longitude (D. Long.). The relationship between Dep. and D. Long. depends upon latitude. In changing Dep. into D. Long., or *vice versa*, use the middle latitude (M. Lat.) which is the latitude midway between the two positions. For example, if one position was in Lat. 54° N and the other was in Lat. 56° N the M. Lat. would be 55° N. Having found, or been given, the D. Long, it can be changed into Dep., and vice versa, by the use of the tables. It will be seen that at the top and bottom of the page Dep. and D. Long. are printed in italics and bracketed, so that by using the degrees at the top or bottom of the page as the M. Lat. one can change D. Long. into Dep. and Dep. into D. Long.

For example, with an M. Lat. of 28° taken from the top of the page (it does not matter whether it is north or south) a D. Long. of 50′ would have

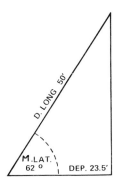

Fig 57   Another diagram showing how traverse tables are used

a corresponding Dep. of 44.1′. In the same M. Lat. a Dep of 57.4′ would have a corresponding D. Long. of 65′. Now turning to the bottom of the page with an M. Lat. of 62° a Dep. of 23.5′ would have a corresponding D. Long of 50′, and a D. Long of 113′ would have a Dep. of 53.1′.

If the M. Lat. does not come to an exact figure one must learn to interpolate by getting the bigger factor, ie D. Lat., Dep., or D. Long. as close as possible in its column, and, if the smaller factor alongside it is not exactly the same as is required, but a little bigger or smaller, interpolate mentally between the quantities on two successive pages. One must try to get answers to the nearest half degree in the course and a half mile in distance.

Plane sailing

When a vessel steers a true course, except north, south, east, or west, the course and distance can be represented on the plane surface of a chart as shown in Fig 58.

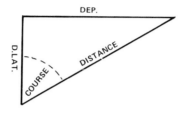

Fig 58   Principle of plane sailing

This is covered up to about 600 miles (the limits of the *Traverse Tables*) but beyond this the curvature of the Earth makes this method inaccurate, and one must use Mercator sailing (which see). Having learned that given a course and distance one can find the D. Lat. and Dep. from the *Traverse Tables*, and if given the D. Lat. and Dep. one could find the course and distance made good. For example, given D. Lat. 79.5′ N and Dep. 42.3′ E the true course and distance made good would be 028° 90 M. With D. Lat. 50.7′ N and Dep. 95.4′ W the true course and distance made good would be 298° 108 M.

Given the following D. Lat. and Dep. what is the course and distance made good?

				Answers		
D. Lat.	21.2′ N	Dep.	11.3′ E	Course 028° N 28° E	Dist.	24 Miles
,,	136.0′ S	,,	72.3′ W	,, 208° S 28° W	,,	154 ,,
,,	79.8′ N	,,	150.1′ W	,, 298° N 62° W	,,	170 ,,
,,	45.5′ S	,,	85.6′ W	,, 242° S 62° W	,,	97 ,,

74

Note very carefully that the course made good (given as a quadrantal reading) are always indicated from the naming of the D. Lat. and Dep., and to avoid making any mistakes it is advisable to name the course in this manner before quoting it in the three figure notation from the figures given at the top and bottom of the page.

Parallel Sailing

This is merely the term used to find the actual distance sailed east and west (or *vice versa*) between two meridians, or to find the departure. It relates to the problem of changing D. Long. into Dep. using the M. Lat. (see Fig 59).

For example, if in Lat. 62° one left Long. 8° W and sailed to a position in Long. 10° W the D. Long. would be 2° × 60 equals 120′ W. In Lat. 62°, 120′ D. Long. equals 56.3′ Dep. which will be seen at the bottom of the page in the extract from the tables given later.

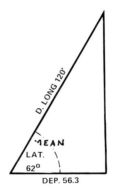

Fig 59   Principle of parallel sailing

Mercator Sailing

In using the *Traverse Tables* it will be seen that the maximum distance given is 600 miles, so that in working problems involving a greater distance than 600 miles one must work to a different formula as illustrated in Fig 60.

The method used is known as Mercator sailing. This is done by using meridional parts. On a Mercator chart the distance between parallels of latitude increase proportionately from the equator towards the poles, and the meridians instead of meeting at the poles are represented by straight lines in a true north and south direction; lying parallel with each other at a constant distance apart. As one leaves the equator 1′ of latitude equals 1′ of longitude, but on reaching Lat. 60° the measure of 1′ of latitude is double the length of 1′ of latitude at the equator.

Meridional parts (MP) simply tabulate the change so that in any latitude one can find exactly the length of a meridian on a Mercator chart, from the equator to the latitude required expressed in minutes of the longitude scale.

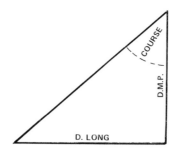

Fig 60   Principle of Mercator sailing

Meridional Parts

The MP for any latitude is the length of the meridian on a Mercator chart measured from the equator to the parallel of latitude and expressed in minutes of the longitude scale. Tables of Meridional Parts are given in *Norie's Tables*, and others, and an extract is given at the back of this book.

Difference of meridional parts (DMP) is the number of minutes of longitude measured along a meridian on a Mercator chart, between any two parallels of latitude. For example, between Lat. 0° (the Equator) and Lat. 0° 01′ the DMP is .99′ which is to all intents 1′, but between Lat. 60° 00′ and Lat. 60° 01′ the DMP is 1.99′, that is almost double. Between Lat. 80° and 80° 01′ the DMP is 5.76′.

Using the extract from the table find the D. Lat. and the Difference of Meridional parts (DMP) between Lat. 52° 53′ N and Lat. 51° 13′ N.

$$
\begin{array}{ll}
\text{Lat. } 52° \ 53' \ \text{N MP} & 3733,47' \\
\text{Lat. } \underline{51° \ 13'} \ \text{N MP} & \underline{3571.25'} \\
\end{array}
$$

$$
\begin{array}{ll}
\phantom{\text{Lat. }}1° \ 40' \ \text{S DMP} & 162.22' \ \text{S} \\
\underline{\times 60} & \\
\end{array}
$$

D. Lat. 100′ S

Again between

Lat. 60° 00′ N and Lat. 59° 05′ N
Lat. 60° 00′ N MP   4507.08′
Lat. 59° 05′ N MP   4398.76′

D. Lat.   55′ S DMP 108.32′ S

Note how the difference of Meridional parts (DMP) is almost double the D. Lat. in the latitude of 60° as mentioned in the earlier paragraph of this section.

Study the following diagram (Fig 61) and the relationship between the various factors in the solution of Mercator sailing problems.

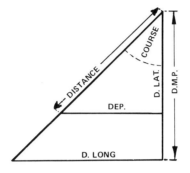

Fig 61   Relation between various factors in solving Mercator sailing problems

The solution can be learned by the use of two simple formulae:

$\dfrac{\text{D. Long.}}{\text{DMP}}$ equals log. tangent of course.

or   Log. D. Long. minus Log. DMP equals Log. tan. co.

and $\dfrac{\text{Dist.}}{\text{D. Lat.}}$ equals log. secant of course.

or   Log. dist. equals log. D. Lat. plus log. sec. co.

The reader may be familiar already with the use of logarithms, and logarithms of trigonometric functions, but if not, extracts from *Norie's Tables* are at the back of the book, which will cover the solution of the following two problems, and a brief explanation is given of their use.

Logarithms

The logarithm of a number is the power to which the base must be raised to produce that number.

This is difficult for a beginner to understand but easier if it is kept firmly in mind that logarithms used in navigation are calculated to base 10, and consist of a whole number called the 'index', and that the index of the logarithm (or log. as it will be called in the future) of any given number is one *less* than the number of figures before the decimal point in the given number.

If 10 is multiplied by 10 it equals 100, or it could be put as 10^2, the small 2 indicating the number of 10's multiplied together. 2 is the 'log.' of 10×10, or the log. of 100. If one multiplies $10 \times 10 \times 10 \times 10$ it would equal 10,000, or 10^4, and the log. of 10,000 is 4. The 100, or 10,000, or whatever figure is used is called the natural number, and the index number of a log. is always one less in value than the total number of figures in the whole natural number, or in other words, those figures which are placed *before* a decimal point. The figures *after* the decimal point are a fraction of the natural number. Do remember that the index depends entirely upon the place occupied by the decimal point. For example:

1892.00 would have an index of 3 because there are 4 figures before the decimal point.

189.20   ,,   ,, ,,   ,, ,,   2 ,,   ,,   ,,   3   ,,

18.92   ,,   ,, ,,   ,, ,,   1 ,,   ,,   ,,   2   ,,

1.89   ,,   ,, ,,   ,, ,,   0 ,,   ,,   ,,   1   ,,

Turning to the extract from *Norie's Tables* at the back of this book headed Logarithms, find the log. for 1892.0. First put down the index which would be 3 (there are four figures before the decimal point) and then look down the column No. (for number) and one will come to 189. Along the top or bottom of the page will be seen the figure 2 and the log. would be expressed as 3.27692.

Now find the log. of

161.4	equals	2.20790
20.96	,,	1.32139
2186.0	,,	3.33965
2.02	,,	0.30535

The next exercise to work is to find the natural number of a given log.

$$
\begin{array}{lll}
2.24080 \text{ equals} & 174.1 \\
1.25334 & ,, & 17.92 \\
3.27738 & ,, & 1894.0 \\
0.32387 & ,, & 2.108 \\
\end{array}
$$

In using logs. the work of multiplying and dividing is made very easy in that instead of multiplying two numbers together one takes out the log. of each number and add the logs. together, and for dividing subtract them. From the log. obtained convert this to a natural number which will give the answer.

Functions of Angles

The term 'functions of angles' relates to the ratio between the three sides of a triangle taken in pairs.

In a right angled plane triangle the relation between any two sides and one of the acute angles (that is any one less than 90°) is called the trigonometrical ratio of that angle as shown in Fig 62. The three sides of a triangle can be arranged in six different pairs, so with six ratios each has a different name. They are sine, tangent, secant, cosine, cotangent and cosecant, and abbreviated they are set down in the following manner which will be used in future: sin, tan, sec, cos, cot, cosec.

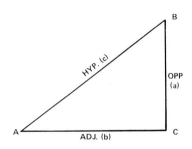

Fig 62   Trigonometrical ratios of triangle

In triangle ABC where we are using angle A then

BC is opposite to the angle (opp.) (a)
AC is adjacent to the angle (adj.) (b)
AB is the hypotenuse   (hyp.) (c)

$$
\text{Sin.} \quad \text{A equals} \quad \frac{BC}{AB} \text{ or } \frac{a}{c}
$$

79

$$\text{Tan.} \quad A \quad ,, \quad \frac{BC}{AC} \quad ,, \quad \frac{a}{b}$$

$$\text{Sec.} \quad A \quad ,, \quad \frac{AB}{AC} \quad ,, \quad \frac{c}{b}$$

$$\text{Cos.} \quad A \quad ,, \quad \frac{AC}{AB} \quad ,, \quad \frac{b}{c}$$

$$\text{Cot.} \quad A \quad ,, \quad \frac{AC}{BC} \quad ,, \quad \frac{b}{a}$$

$$\text{Cosec.} \, A \quad ,, \quad \frac{AB}{BC} \quad ,, \quad \frac{c}{a}$$

When using the *Traverse Tables* one is solving right-angled plane triangles by inspection, and the hypotenuse is the Dist., the adjacent side is the D. Lat., and the opposite side is the Dep., and the angles, the course.

One must learn to find and use logarithmic values of these sines, cosecants, tangents, *etc* from the extract from *Norie's Tables* at the back of this book. The page is headed Logs. of Trig. Functions, and is headed in the order (from left to right) sine, cosec, tan, cotan, secant, and cosine. The degrees are marked in the top left hand corner, and the minutes are in the column immediately below the figure of degrees (46° in this case). When using decimals of a minute the columns headed 'Parts' give the figure either to add to *or* subtract from the value obtained for that particular minute (one must look at the next greater minute to see whether to add or subtract).

For example, if one requires the log. sine of 46° 10.0′ look down the sine column and opposite 10.0′ will be found the figure of 9.85815. For the cotan. of 46° 59.0′ the figure would be 9.96991. Now, if the secant is required for 46° 29.4′ one would get 10.16206 plus the parts for .4 which is 5, so the complete answer would be 10.16206 plus 5 equals 10.16211.

Find the following functions of angles from the tables.

Cosine of 46° 3.8′ equals 9.84138 minus 11 gives 9.84127
Tan. ,, 46° 29.0′ ,, 10.02250
Sine ,, 46° 48.2′ ,, 9.86271 plus 2 gives 9.86273

For angles of less than 4° the changes are so large that the minute column under the degrees is divided into .2′ for each minute.

Examinees may be required to convert functions to angles, and to do this the above process is reversed. First you find the appropriate column in the tables for the function nearest to the one you require, looking carefully at the index number of the log.

To find the angle of which 10.15875 is the secant the answer would be 46° 04.0′, or of which 9.86060 is the sine, it would be 46° 30.0′ (9.86056) plus 4 for which the decimal part would be .3, so the answer would be 46° 30.3′.

Find the angles of the following.

Cosec. 10.13980 equals 46° 27.0′
Cotan. 9.97350 ,, 46° 44.0′ plus .8 gives 46° 44.8′
Secant 10.16492 ,, 46° 50.0′ ,, .4 ,, 46° 50.4′

Having now learned to use these tables the solution of two Mercator sailing problems can be given.

1. To find the course and distance from A in Lat. 41° 36′ N Long. 5° 01′ W to B in Lat. 60° 39′ N long. 38° 00′ W.

Remember the formula $\dfrac{\text{D. Long.}}{\text{DMP}}$ equals tangent of course.

or

Log. D. Long. minus log. DMP equals log. tan. co.
and

$$\frac{\text{Dist.}}{\text{D. Lat.}} \text{ equals secant of course}$$

or

Log. Dist. equals log. D. Lat. plus log. sec. co.
Set work out as follows:

| Lat. A 41° 36′ N | MP 2733.97 | Long. A 5° 01′ W |
| Lat. B 60° 39′ N | MP 4585.72 | Long. B 38° 00′ W |

D. Lat. 19° 03′ N	DMP 1851.75	D. Long. 32° 59′ W
× 60		× 60
1143′ N		1979′W

D. Long. 1979′ Log. 3.29645   D. Lat.   1143′   Log.   3.05805
DMP   1852′ Log. 3.26764 + Co. N 46° 54′ W   Sec.   10.16541

Log. tan. co. 10.02881   Log. Dist.   3.22346

Answers: Course N 46° 54′ W   Distance 1673 miles.   (See Fig 63)

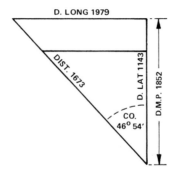

Fig 63   Diagrammatic solution of course and distance problem

2.   A vessel steered 046° T for 1560 miles. If it sailed from Lat. 42° 10′ N Long. 37° 50′ W what was its position at the end of the run?

$$\text{To find D. Lat. } \frac{\text{D. Lat.}}{\text{Dist.}} \text{ equals cos. co.}$$

or

Log D. Lat. equals Log. Dist. $+$ log cos. co.

$$\text{To find D. Long. } \frac{\text{D. Long.}}{\text{DMP}} \text{ equals tan. co.}$$

or

Log. D. Long. equals Log. DMP $+$ log tan. co.

Dist.   1560 M Log.   3.19313   Lat. left 42° 10′ N   MP   2779.47

Co. N   46° E Cos.   <u>9.84177</u>   D. Lat.   <u>18° 04′ N</u>

       D. Lat. log   3.03490   Lat. in   60° 14′ N   MP   <u>4535.13</u>

                                                   DMP 1755.66

D. Lat. 1084′ N
equals 18° 04′ N

DMP 1756   Log.   3.24452   Long. left 37° 50′ W

Co. N 46° E   Tan.   <u>10.01516</u>   D. Long.   <u>30° 18′ E</u>

    D. Long. Log.   3.25968   Long. in   7° 32′ W

D. Long. 1818 E
equals 30° 18′ E

It will be seen that when adding together the log. of a number and the log. of a trigonometrical function discard the 10 because the index of a log.

sine, tangent, etc. is increased by 10 to avoid a negative index. Similarly, when subtracting two logs. to obtain the log. sine, tangent, etc. add 10 to the index of the log. of the number from that which one is subtracting (Fig 64). For example:

Log. sec. 46° 24.00'   10.16139
Plus log. 110   2.04139

equals log.   2.20278

Answer: 159.5   (The 10 is discarded from the index
                  which otherwise would have been 12)

and by subtracting from the log. of 114.0   2.05691

the log. of   109.2   2.03822

one obtains the log. tan.   10.01869
which equals 46° 14.0'

(10 is added to the index)

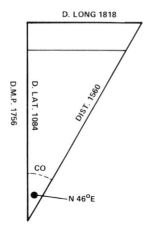

Fig 64   Diagram showing solution of calculation of position at end of run

Chapter 6   Nautical Astronomy

INTRODUCTION — DEFINITIONS — FIGURE DRAWING — USE OF
CHRONOMETER — SUN'S MERIDIAN PASSAGE — GREENWICH
MEAN TIME — USE OF THE NAUTICAL ALMANAC — FINDING
DEVIATION BY A BEARING OF THE POLE STAR, SUN ON THE MERIDIAN,
AMPLITUDE OR TRUE AZIMUTH.

Introduction

When out of sight of land a ship's position can be found by means of
taking observations of the sun. As described in Chapter 4, which deals
with the use of the sextant, it is possible by measuring the vertical angle
of the sun to find a position at sea to a high degree of accuracy. By taking
bearings of the sun the deviation of the compass can be found at any time.

Definitions

A number of terms and definitions have been given already in Chapters 2
and 4 which apply to the work involved in nautical astronomy. The
remaining definitions that an examinee may be required to know are
given in this chapter. They must be memorised and their exact meaning
known for use in working various problems that may be set.

Apparent Time

This is measured by the true (or apparent) sun. Owing to the earth's
varying distance from the sun as it progresses on its orbit around the sun,
the time at which the sun's centre crosses the meridian of a fixed observer
between two successive transits (that is at apparent noon on two successive
days) alters slightly at each day, either earlier or later. It amounts to only
a few seconds each day but accumulates as the days pass until it amounts
to nearly $16\frac{1}{2}$ minutes.

Mean Time

As it would be impossible to construct a mechanical timepiece to keep
pace with the true (or apparent) sun during those periods throughout the
year when it slowed down or speeded up, the astronomers have devised a
mean sun, which moves along the equinoctial at a uniform rate equal to
the *average* rate of the true sun on the ecliptic. This is the time in general
use for all civil purposes, and a day is considered to last exactly 24 hours.
The equinoctal is the celestial equator.

Equation of Time

This is the difference in minutes and seconds between mean time and apparent time. As mentioned above it changes very slowly from day to day, and the equation of time is given for 0000 hours and 1200 hours each day in the *Nautical Almanac*. The reader should look at the extract of this publication at the back of the book where it will be seen in the bottom right hand corner of the page.

Prime Vertical

The vertical great circle (Fig 65) is that which passes through the observer's zenith and cuts the horizon at its true east and west points.

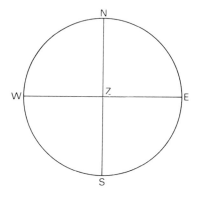

Fig 65   Prime vertical. NESW shows the rational horizon. Z is the observer's zenith. WZE is the prime vertical NZS is the observer's meridian

Azimuth

This is the angle at the observer's zenith between a celestial meridian drawn through the observer's position and a vertical circle through the centre of the sun, measured from the elevated pole as illustrated in Fig 66.

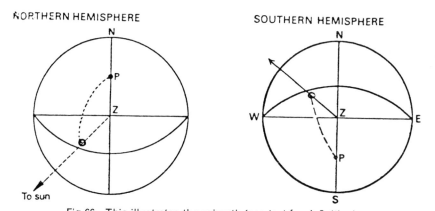

Fig 66   This illustrates the azimuth (see text for definition)

Celestial Concave

To understand nautical astronomy imagine an observer on the earth looking outward into space and seeing the celestial bodies (sun, moon, stars, planets, *etc*) projected on the interior concave surface of the *celestial sphere*. If a line is drawn outwards from the centre of the earth (Fig 67) to the celestial body the point where the line passes through the surface of the earth is the body's geographical position Y at that moment of time.

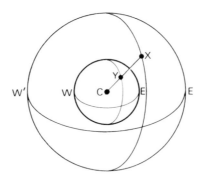

Fig 67   This figure explains the concept of the celestial concave in relation to the earth and a celestial body (see text). X marks the celestial position of the body. WE is the earth's equator. Y is the geographical position. WE marks the celestial equator and C indicates the centre of the earth

Celestial Equator

The equator is a great circle drawn around the earth, midway between the poles. If this great circle is extended to the celestial sphere it becomes the celestial equator, or equinoctial as it is called.

Ecliptic

It has been shown already that the axis of the earth is not at right angles to the path of its oribit around the sun, but that it is tilted about $23\frac{1}{2}^\circ$ from the vertical. The orbit it makes is not a true circle but is slightly elliptical in shape, and as a result of this the length of each day varies. The varying seasons of the year are caused by the changing declination of the sun.

To an observer on the earth the sun *appears* to rotate in an orbit about the earth. This orbit, when projected on to the celestial sphere is known as the *ecliptic* (Fig 68).

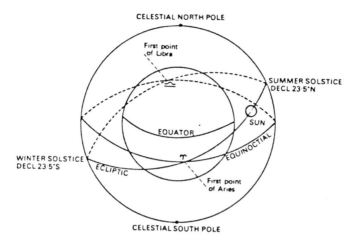

Fig 68   Explains the ecliptic showing the apparent orbit of the sun round the earth as projected on the celestial sphere

Local Hour Angle of the Sun (LHA)

This is the angle at the pole between the meridian of the observer and a meridian through the geographical position of the sun's centre, and it is always measured westward from the observer's meridian as in Fig 69.

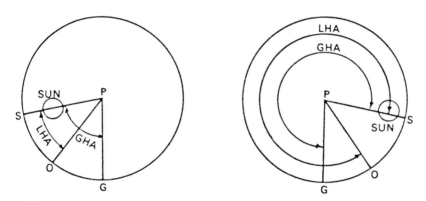

Fig 69   Shows the local hour angle of the sun normally recorded as LHA

P. pole. PG Greenwich meridian, PO observer's meridian
PS meridian of the sun's geographical position. Angle GPS equals GHA and Angle OPS equals LHA.

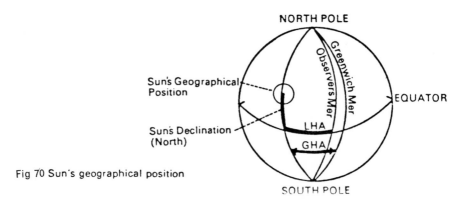

Fig 70 Sun's geographical position

Remember two simple formulae:

Long. W, Greenwich hour angle best
Long. E, Greenwich hour angle least
therefore LHA equals GHA minus west long
therefore LHA equals GHA plus east long.

Geographical Position of a Heavenly Body

This is the point on the surface of the earth over which it would be directly overhead, and the latitude would correspond to its declination, and its longitude would equal the Greenwich hour angle at that instant (see drawing of celestial concave on page 86).

Position Line

In nautical astronomy the position line is a small portion of an arc of a small circle the centre of which is the geographical position of the sun the radius being equal to the zenith distance of the sun. As the portion of the arc is so small it can be drawn as a straight line at right angles to the sun's true bearing.

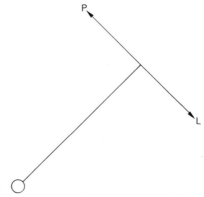

Fig 71 Records the position line (PL) of the observer in relation to the sun's true bearing

As described previously the visual bearing of a shore light, or point of land, is a position line at some point on which is the observer's position (Fig 71).

Declination

Due to the earth tilting at an angle of approximately $23\frac{1}{2}°$ as it progresses on its orbit around the sun the 'latitude' of the sun varies from day to day, moving from a maximum of 23° 26.6′ N on June 21, to a maximum of 23° 26.6′ S on December 21. The correct name given to this 'latitude' is the Sun's declination. If, for instance, one were in Lat. 18° 20′ N and on that day the sun's declination was given as 18° 20′ N it would be directly over head at NOON. Declination is given for every hour (GMT) of each day in the *Nautical Almanac*. Later is given the definition of a celestial meridian and the celestial equator, so that in astronomical terms declination can be described as the arc of a celestial meridian between the celestial equator and a small circle drawn parallel to this equator through the sun. If the word 'celestial' is removed from this definition it will be seen that it corresponds exactly to the definition of latitude, and therefore, the declination of the sun corresponds to terrestrial latitude (Fig 72).

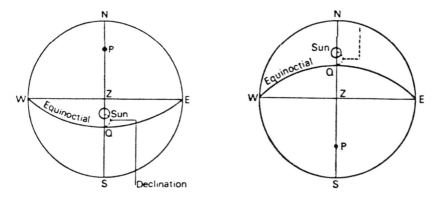

Fig 72 Demonstrates the declination of the sun. Z is the observer's zenith, WQE is the equinoctial. O is the sun on the meridian. P is the elevated pole. QO is the sun's declination north. NESW is the rational horizon. PO is the polar distance

Polar Distance (PD)

This is the arc of a celestial meridian between a small circle drawn through a celestial body (such as the sun) and if the Decl. is north and one is in a North Lat. the PD equals 90°–Decl., but if the Decl. is South the PD equals 90° plus the Decl.

Greenwich Hour Angle of the Sun (GHA)

This is the angle between the meridian of Greenwich and a meridian drawn through the geographical position of the sun's centre at any specific instant of time. It is always measured westward from Greenwich and is given for every hour of GMT in the *Nautical Almanac*.

Such is the precise degree of accuracy of the passage of the earth around the sun that it takes exactly 365 days 5 hours 48 minutes 46 seconds to complete its orbit.

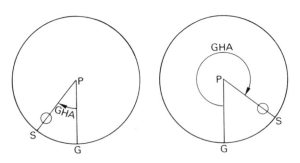

Fig 73 These two figures demonstrate finding the Greenwich hour angle of the sun (GHA). P is the north poles. O is the sun. PG is the Greenwich meridian. PS is the meridian of the sun. The angle GPS or the arc GS equals the Greenwich hour angle (GHA)

Figure Drawing

To understand the actual working of the problems in nautical astronomy it is always of great assistance to draw a figure to show how one arrives at the answer. The method used is called the stereographic projection of the celestial sphere on the plane of the rational horizon (see Fig 74a). A circle is drawn which represents the rational horizon as a great circle on the celestial concave, every point of which is 90° from the zenith of the observer. In other words, if one were able to project oneself into the celestial sphere, rather like an astronaut who has travelled beyond the sun, one would reach a position directly above (the zenith) and would be looking down upon the earth from this tremendous height, so that all the details of the celestial sphere and the sun could be drawn within this circle to represent the data required in solving problems in nautical astronomy.

To draw a sketch when the sun is not on the meridian would appear as Fig 74b opposite (sun west of meridian).

X is the position of the sun on its daily path d′ O d from east to west equal to its declination (north).

The positions of d d′ can be found by taking out the amplitude of the sun which we shall work out later in this chapter.

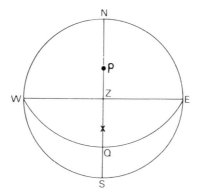

Fig 74a Stereographic projection to check working problems in nautical astronomy. In Figure a the circle NESW represents the rational horizon. Z is the observer's zenith. NZS is the observer's meridian. WZE is the prime vertical. P the elevated pole — *ie* the pole visible from your latitude, north in this case. X is the sun on your meridian. WQE is the equinoctial. QX is the sun's declination (north). QZ latitude. PX polar distance. SX true altitude. ZX zenith distance

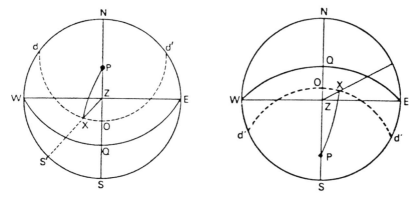

Fig 74b This sketch illustrates the position when the sun is not on the meridian and the sun is west of the meridian and east of the meridan

Angle ZPX is the sun's hour angle.
Angle PZX is the azimuth of the sun.

When drawing these figures use a radius of 9/10 ins. so that every 10° of arc can be plotted as 1/10 ins. For example if you wished to plot an altitude of the sun on the meridian as 50° S you would measure 5/10 ins. from S on the meridian.

To find the centre of the circle joining PX, first find by trial the position on the meridian NZS whereby a circle can be drawn passing through the points WPE and call it Y. From this point extend a line at right angles to ZS which line is called a 'locus', and the centre of all meridians through P

91

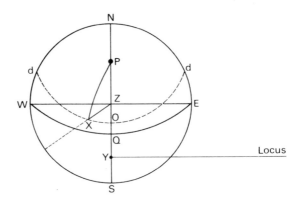

Fig 75a This illustrates further stage in the exercise fixing the centre of the meridian somewhere on the locus

will be somewhere on the locus. By trial find the point on the locus where the part of the great circle PX can be drawn.

If the sun was to the east of the observer's meridian the locus would be drawn to the westward using exactly the same methods. This is illustrated below with the declination shown as south.

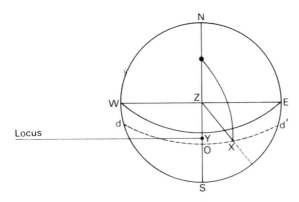

Fig 75b Illustrates the position with the sun to the east of the meridian (see text)

Use of Chronometer

GMT (which is essential for position finding when out of sight of land) is kept on board by a chronometer, which is an extremely accurately constructed clock, specially designed and compensated for changes in temperature. A chronometer is slung in gimbals like a compass and kept in a box with a glass lid which is opened only when the chronometer has to be wound, which should be daily, and at the same time each day.

If a chronometer is not on board an accurate deck watch, or chronometer watch, will be sufficient, provided its accuracy is checked daily by means of time signals received by radio. A chronometer watch is wound by means of the winding stem, like an ordinary watch.

To wind a chronometer turn the instrument gently upside down on its gimbals and push aside a small dust cap beneath which is the aperture to take the winding key. It is wound by turning in an anti-clockwise direction of between 7 and 8 turns. A small hand on the dial of the chronometer indicates when it is fully wound.

Never alter the hands of a chronometer. Any error should be noted in a chronometer rate book each time the chronometer is checked with a broadcast time signal. The error is found by comparing the chronometer reading with the time signal. These radio time signals can be obtained from all over the world, the details of such transmissions being obtained from Vol 5 of the *Admiralty List of Radio Signals*. If the BBC time signals are used remember it is the last dot which indicates the exact hour.

From the time signals received it is possible to calculate the daily rate to find whether the chronometer is losing or gaining, and by multiplying the number of days by the daily rate one can calculate the *accumulated rate* to apply to the last error obtained by a time signal. For example, if on October 3 the chronometer was 2 mins. 6 sec. slow of GMT and the daily rate was 2 seconds losing, what would be the error on October 10?

Daily rate 2 sec. losing	Error Oct. 3 2 m 6 s slow
No. of days 7_____	Accum. rate
Accum. rate 14 secs. losing	losing plus 14 s
	Error Oct. 10 2 m 20 s slow

If it is found that the error is becoming too large the chronometer should be returned to the maker's agent for it to be re-set and corrected.

Sun's Meridian Passage

On the bottom right hand side of the page from the *Nautical Almanac* covering May 3, 4, 5, will be seen the time of the sun's meridian passage headed Mer. Pass. This is the Greenwich mean time at which the sun will cross the Greenwich meridian on that day, and is also the local or ship's mean time in longitude east or west of Greenwich.

Greenwich Mean Time (GMT)

This is taken as standard all over the world, each day being divided into 24 hours, and the length of the hour never alters. It relates to the mean time kept at the meridian of Greenwich 0° and is the time kept by the chronometer as described earlier. All data in the *Nautical Almanac* relates to GMT and will be used in the problems candidates will have to solve. The GMT must always be obtained to enable one to go to the *Nautical Almanac* and extract the data required.

In order to find the GMT one must learn to convert arc into time. Longitude is expressed in arc, that is in degrees, minutes and seconds, and at the back of this book is an extract from the *Nautical Almanac* which shows how to convert arc into time. For example, in longitude 15° 30' West is equal to 1 hour 2 minutes in time. From the conversion of arc to time table it will be seen that 15° equals 1 hour 00 minutes, and on the right hand side of the page 30' equals 2 minutes.

Longitude west Greenwich time best
Longitude east Greenwich time least

Always remember that as the earth revolves from west to east a place situated east of the Greenwich meridian 0° would be ahead in time of Greenwich, and the GMT would be less. If west of Greenwich then the GMT would be greater.

A chronometer will give the GMT without any calculations being required, but in most examination problems in nautical astronomy it is necessary to calculate what the GMT will be in order to take out the necessary data from the *Nautical Almanac*.

Before proceeding further work a few more examples of conversion of arc to time (work to the nearest minute).

1. Long. 8° 55' West   2. Long. 12° 48' W.   3. Long. 6° 22' East
4. Long. 10° 55' W.   5. Long. 9° 41' E.   6. Long. 23° 59' W.
7. Long. 10° 02' W.   8. Long. 5° 56' E.

Answers

	Hours	Minutes		Hours	Minutes		Hours	Minutes
1.	0	36	2.	0	51	3.	0	25
4.	0	44	5.	0	39	6.	1	36
7.	0	40	8.	0	24			

GMT is necessary to observe a meridian altitude of the sun to find latitude, that is when the sun is on the observer's meridian and bearing exactly south; that is when it has reached its maximum altitude. GMT is also required at the time the sun is rising or setting in order to work an amplitude.

On the bottom right hand side of the page from the *Nautical Almanac* dated May 3, 4 and 5, will be seen the time of the sun's meridian passage, as indicated already. This is the GMT at which the sun crosses the

Greenwich meridian. If one is on that meridian at that instant of time a chronometer which is keeping GMT, would show the mean time that the sun was on the meridian. If east or west of Greenwich the sun would be on the observer's meridian at the same *local or ship's mean time*, but it would be necessary to apply the longitude in time to obtain the GMT (see Fig 76).

Below are some examples showing local mean time of the Mer. Pass. and the longitude, with the resultant GMT of Mer. Pass.

1.	Longitude	12° W	LMT Mer. Pass			1210
2.	,,	10° 30′ E	,,	,,	,,	1158
3.	,,	4° 45′ W	,,	,,	,,	1200
4.	,,	2° 15′ E	,,	,,	,,	1145

Answers
1. GMT 1258   2. GMT 1116   3. GMT 1219   4. GMT 1136

No. 1 Illustrated
PG   Greenwich Meridian
PO   Observer's Meridian
OG   Arc. Long. 12° West equals 48 min. in Time
X   Sun on Observer's Meridian
Mer. Pass. 1210 plus 48 min. equals 1258 GMT.

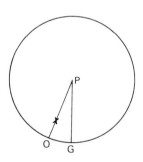

Fig 76 Securing local or ship's mean time.

Use of Nautical Almanac

Extracts from the *Nautical Almanac* are given at the back of this book sufficient to enable the reader to work the examples given in this, and the following chapter. Remember that all times in the almanac relate to GMT.

Similar data can be obtained from various nautical publications which are published annually.

For the purpose of the Department of Trade Examinations candidates are required to know how to use the data contained in the *Nautical Almanac*, and this includes *altitude correction tables, daily ephemerides of the sun* (Greenwich hour angle, declination, Meridian passage, equation of time, sunrise, sunset, twilight), conversion of arc to time, increments and corrections, tables for interpolating sunrise, *etc.*

Practice being able to find the correct declination which is required for working the various problems in nautical astronomy.

On the left hand side of the page for May 3, 4 and 5, is the column headed sun, and a sub-heading of Dec. which is an abbreviation of declination. If the Dec. for May 4 at 1800 hours is required (always use the four figure notation in describing time, the first two figures relating to the hours, and the last two are minutes) it would give 16° 08.9′ N. This means that the sun is 16° 08.9′ N of the equinoctial (or celestial equator) and correspondingly the equator, so that in that latitude the sun would be directly overhead at noon.

Look out the Dec. for the following dates and times:

1. May 4 2200 GMT equals 16° 11.8′ N
2. May 3 0600   ,,   ,,   15° 42.8′ N
3. May 5 0100   ,,   ,,   16° 13.9′ N
4. May 3 1200   ,,   ,,   15° 47.2′ N

These have been taken for an even number of hours, but obviously this would happen only on very rare occasions, and normally the GMT would be in between the exact hour.

At the foot of the Dec. column will be seen a small 'd' with 0.7 beside it. That shows by how much the Dec. changes each hour on the average for that page. There are tables (increments and corrections) which can give quickly and easily the change that takes place between the exact hours for so many minutes. An extract is shown at the back of the book for 44 mins. and 45 mins. and if, for example, it is required to know the correction to apply to the Dec. on May 4 for 16 hrs. 44 mins. (1644) it would be 16° 07.5′ N, and with the hourly change of 0.7′ the table gives a correction of 0.5′. When applying this correction always look at the next hour to see whether the Dec. is increasing or decreasing, and as in this case it is increasing the 0.5′ is added, making the correct Dec. for 1644 GMT as 16° 08.0′ N.

Finding Deviation of the Compass by a Bearing of the Pole Star, Sun on the Meridian, Amplitude.

The Pole star is situated close to the true north pole of the heavens, and a bearing of it can be taken at any time during the hours of darkness to check deviation on any heading. Although at certain periods of the night it might be approximately 1° from the true north it gives a good approximate check on deviation.

The procedure is simple, because the bearing of the pole star is always taken as 000° (or 360°) so that when a compass bearing is taken of it the difference will be the error of the compass, and by applying the variation to this the deviation can be obtained.
For example:

True bearing of pole star	000°	True bearing of pole star	000°
Compass bearing	010°	Compass bearing	356°
Error	10°W	Error	4°E
Var.	8°W	Var.	2°W
Dev.	2°W	Dev.	6°E

Always remember that error W. compass is best, and error E. compass is least. Having found the error be very careful to name the Dev. correctly when applying the Var. As an independent check apply the sum of the Var. and Dev. to make sure that it comes to the same sum as the error (same names add, different names subtract). If the Dev. is named incorrectly it will show immediately.

The pole star can be found by looking towards the north at an elevation equal to the observer's latitude. The following sketch of the Great Bear will assist (Fig 77).

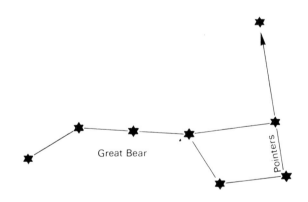

Fig 77 Finding the Pole Star

97

Sun on the Meridian

When the sun is on the observer's meridian at noon in northern latitudes it will always bear true south, or 180°. If a bearing is taken of it at this moment the difference between the true bearing and the bearing observed by compass would give the compass error. As in the previous problem by applying the Var. to the error the Dev. can be found (Fig. 78).

For example:

True bearing of the sun	180°		True bearing of the Sun	180°
Compass bearing	188°		Compass bearing	174°
Error	8°W		Error	6°E
Var.	7°W		Var.	1°W
Dev.	1°W		Dev.	7°E

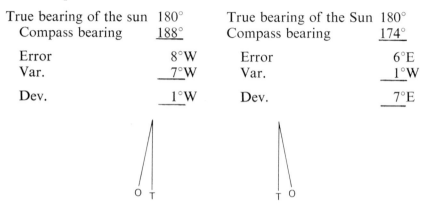

Fig 78 Naming the error when the Sun is on the meridian

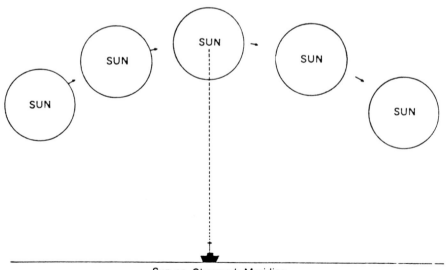

Sun on Observer's Meridian
Reaches Maximum Altitude.

Fig 79 Sun on the meridian, maximum altitude

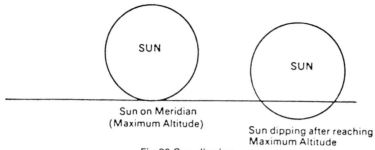

Sun on Meridian
(Maximum Altitude)

Sun dipping after reaching
Maximum Altitude

Fig 80 Sun dipping

Amplitude

When a bearing is taken of the sun rising or setting it is called an amplitude, and it is actually the angle between the east point and the body when rising, and the west point and the body when setting. As the refraction of the sun is considerable at these times, to take a bearing of the sun's centre as it rises or sets it must be taken when the lower limb of the sun is half its diameter above the horizon as in Fig 81.

Horizon

Fig 81 Amplitude—setting or rising sun—must register half a diameter above the horizon

When working an amplitude problem examinees are given the latitude and longitude, both of which are required to complete the solving of the problem. Sometimes candidates are given the GMT or the LMT at the time of the sunrise or sunset, but if not, one can obtain the local mean time of sunrise or sunset for the relative latitude from the *Nautical Almanac*. It will be seen from the page at the back of the book covering May 3, 4, 5, that the times given differ according to latitude. If the latitude does not correspond exactly to that given in the tables it must be interpolated to get the correct time of the sunrise or sunset. It is quite a simple exercise to do this, but if any difficulty is encountered there is a table in the *Nautical Almanac* to assist in the interpolation.

Having found GMT by applying the longitude in time to the LMT of sunrise or sunset (given in the problem or found from the *Nautical Almanac*) then enter the almanac to find the correct declination.

With the latitude and the correct Dec. turn to the true amplitude table, given at the back of this book, and using the Dec. at the top of the page and the Lat. at the side, read the figure where these two factors meet in the relevant column. This figure gives the true amplitude of the sun

99

when it is rising or setting, and if this is converted to the true bearing and compared with the compass bearing taken at that time, the difference will be the error of the compass. By applying the Var. to this the Dev. will be found as in the previous problems.

The most important point to remember in the working of an amplitude problem is the actual naming of the true amplitude obtained from the tables. The rule is to name it *east* if rising, and *west* if setting, followed by the *same* name as the Declination. For example, if it is found that the Sun's True Amplitude from the table was 24° when rising, and the Dec. was North for that day, the bearing would be given as E 24° N. Again, if the true amplitude of the sun's setting was 14° and the Dec. was south, it would be given as W 14° S (Fig 82).

In order to compare the True bearing with the Compass bearing it is necessary to change the figure obtained from the amplitude table into the three figure notation. For example, E 24° N would be that number of degrees to the north of east, and as east is 090° it would be 090°–24° equals 066° which is expressed in the correct three figure notation.

Study the following examples:

$$
\begin{array}{llllllll}
\text{W } 10° \text{ N} & \text{equals} & 270° & \text{plus} & 10° & \text{equals} & 280° \\
\text{W } 20° \text{ S} & ,, & 270° & \text{minus} & 20° & ,, & 250° \\
\text{E } 10° \text{ S} & ,, & 090° & \text{plus} & 10° & ,, & 100° \\
\text{E } 15° \text{ N} & ,, & 090° & \text{minus} & 15° & ,, & 075° \\
\end{array}
$$

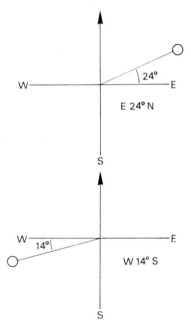

Fig 82   Diagram to name true amplitude (see text)

Candidates will be required to work these problems to decimal points, and the two following examples are given to show how this is done.

On May 4 in Estimated position 50° N 5° W at 0600 GMT the sun rose bearing 072° by compass. Find the compass error and deviation for the direction of the ship's head, the Var. being 4° W.

May 4 1972

G.M.T.	0600	True bearing from amplitude table 25.4
Dec.	16° 00.3′ N	Named *east* because rising and *north* because
Lat.	50° N	the *declination* is *north*

True	Amp. E 25.4°N	Subtract from
	90°	

Brg.	064.6°	True
Var.	4°W	

	068.6°	Mag.
	072.0	Comp.

Dev.	3.4°W	

Variation is *added* because it is west, and as the compass bearing is greater (best) than the Magnetic bearing, the Deviation must be west (error west, compass best).

Now for an example where the GMT is not given and has to be found. On May 4 in E.P. 52° N 10° W the sun set bearing 291° by compass. Find the compass error and the deviation for the direction of the ship's head, the variation being 9° W.

May 4 1972

Mean time sunset Lat. 52° N	1930	True ampl. W	26.6°N	*West* because
Long. in Time 10° W plus	40		270°	setting and
GMT	2010	Brg.	296.6°T	*North* as declina-
Dec. 16° 10.4 N		Var.	9.0°W	tion
Lat. 52° N.		Mag. bear.	305.6°M	
		Comp. bear.	291　C	
		Deviation	14.6°E	

In the examination candidates are required to work to the nearest 0.5 of a degree, and it needs interpolation between the degrees of *latitude* and *declination* to attain this. By careful working, one can obtain an answer to the nearest decimal point, and this degree of accuracy should be attempted (Fig. 83).

For example, in lat. 49° N dec. 16½ N the correct figure required would be between 16° and 17°, or 24.9° and 26.5°, as will be seen from the tables. To interpolate here one would subtract one from the other and divide by two, adding the result to the *lesser* figure.

$$\begin{array}{r} 26.5° \\ 24.9° \\ \hline \text{Divide by 2 } \quad 1.6° \\ \hline 0.8° \end{array}$$

24.9° plus .8° equals 25.7°

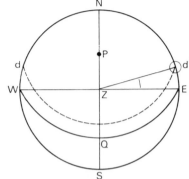

Fig 83   Test diagram for amplitude (see detail in following text)

Diagram showing the sun *rising* in Lat. 50° N with Dec. 10° N.

NESW	Rational horizon
Z	Observer's zenith
NS	Meridian through this point
P	Elevated pole
WQE	Equinoctial
dd′	Circle of declination
O	Sun rising
Angle EZO	True amplitude

Finding Deviation of the Compass by True Azimuth of the Sun

The azimuth of the sun is the angle at the observer's zenith between the observer's meridian and a vertical circle through the centre of the sun measured from the elevated pole, and to find this the value of the following three factors are required.

1. Latitude   2. Declination of the sun   3. The local hour angle of the sun

NESW	Rational horizon
Z	Observer's zenith
NS	Meridian through this point
P	Elevated pole
WQE	Equinoctial
X	Sun's position on its circle of declination dd′
RT	Vertical circle through X
Angle PZX	is the azimuth of X
Angle ZPX	is the hour angle

102

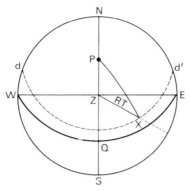

Fig 84 Finding compass deviation by the sun's true azimuth (for detail see text)

The Lat. would be given in an examination problem, but in practice one's own latitude would be known. The Dec. would be obtained from the *Nautical Almanac* as learned previously.

Candidates are required to know their GMT which could be given in the problem, or obtained from their chronometer at the moment of time a compass bearing of the sun is taken. If LMT is given then by applying the Long. (in time) to this the GMT can be found.

Having obtained GMT by any of the above methods turn to the *Nautical Almanac* and find the Greenwich hour angle (GHA) of the sun for that GMT. Look at the page covering May 3, 4 and 5, and under the column headed SUN will be seen the GHA given for every *hour*. In the tables increments and corrections are given for the change in the GHA for every minute and second between each hour, so that if it is required to know the GHA for May 3 at 19 hr. 45 min. 30 sec. GMT it would be taken out for 19 hrs. which equals 105° 48.1′. For 45 min. 30 sec. the change is 11° 22.5′ which is always *added* to the GHA for the hour.

GHA for 19 hrs.	105° 48.1′
Correction for 45 m 30 s	$+11°$ 22.5′
Correct GHA for 19–45–30	117° 10.6′

Having found the correct GHA for the GMT apply to it the Long. (in arc) to obtain the local hour angle (LHA). How this is done is explained in the definition of LHA earlier in this chapter, but to make sure the formula is firmly in mind it is repeated here.

LHA equals GHA minus west long.
LHA ,,   ,, plus east long.

Two examples are given here:
GHA 315° 47.5′ in longitude 8° 42′ W the LHA would be 315° 47.5′ minus 8° 42′ equals 307° 5.5′.

GHA 60° 47.9′ in longitude 5° 10′ E the LHA would be 60° 47.9′ plus 5° 10′ equals 65° 57.9′.

With the three factors Lat. Dec. and LHA turn now to the ABC tables which are published in *Norie's*, an extract of which is at the back of this book.

Table A is entered with the LHA at the top or bottom of the page, and the Lat. at the sides. The rules for naming the figure obtained from this table are set out in each margin of the page. For example, in Lat. 50° N LHA 46° W from Table A a figure of 1.15, would be obtained and it would be named *south* (opposite to the Lat.). From Table B with the same LHA and a Dec. of 15° N would give .37, which is named *north*, the same name as the Dec.

Set these figures out as follows: A.   1.15 S
                                    B.   −.37 N

                                    C.   0.78 S

The rule is: same names add, different names subtract, to find the value of C. If in any doubt about how to name C instructions are given clearly at the bottom of the page in Table C.

With C in our example equalling 0.78 S taken from either the top or the bottom of the page, and with Lat. 50° N we obtain the figure of 63.4° (in using the ABC Tables if the exact Lat. Dec. and LHA, or factor C are not given, one must interpolate).

The true azimuth takes the combined names of C and the hour angle so that 63.4° would become S 63.4° W, because C is named south and the hour angle is west. Changing this into the three figure notation it becomes 243.4°. If, in this case, the compass bearing of the sun had been 255° and the Var. 9° W the error of the compass and the Dev. would have been found as follows.

True bearing	243.4°
Compass bearing	255°
Error	11.6° W
Var.	9.0° W
Dev.	2.6° W

In naming the hour angle remember that if it exceeds 180° it should be subtracted from 360° and named *east* to obtain the correct naming of the azimuth.

Another way to remember this is as follows: the hour angle is west if between 0° and 180°, and east between 180° and 360°. Another important point to remember is that the figure taken from the ABC tables called the azimuth is really a *true* bearing in quadrantal notation.

Chapter 7   Nautical Astronomy (*continued*)

Finding Latitude by Meridian Altitude of the Sun

Having studied in previous chapters the terms and definitions required to solve this problem, one must now learn how to apply them. Using the data from the *Nautical Almanac* at the back of this book find the latitude given the following details:

On May 4 1972 in DR position 49° 10′ N 8° 00′ W the sextant altitude of the sun's lower limb, when on the meridian, was 56° 48′ bearing south. Height of eye 17 feet, index error 1′ on the arc.

Set work out in the following order and get into the habit of putting down all the necessary details in a methodical manner so that at the examinations the examiner can follow the working, in order to see how the answer was obtained.

May 4 1972

LMT Mer. Pass	1157	Sextant alt.	56° 48′ South
Long. in time W plus	32	Index error	−1.0
GMT	1229	Observed alt.	56° 47.0
		Corr. for dip.	− 4.0
Declination 1200	16° 04.6N	Apparent alt.	56° 43.0
'd' 0.7 correction plus	0.3	Correction plus	15.3
Correct declination	16° 04.9N	True alt.	56° 58.3′ S
			90°
		Zenith distance	33° 01.7 N
		Declination	16° 04.9 N
		Latitude	49° 6.6′ N

Notes on the working.

(a) Mer. Pass. was 1157, and adding Long. in time (Long. west Greenwich time best) gives GMT 1229.

105

(b)   Dec. for 1200 hours GMT is 16° 04.6′ N and for the odd 29 minutes as little 'd' was 0.7, and as 29 min. equals 0.5 of an hour, 0.7 × 0.5 equals 0.35 (say 0.3) added as the Dec. is increasing.

(c)   As the index error was 1′ on the arc subtract this from the sextant altitude. For the Dip correction look in the column headed DIP and with height of eye 17′ the correction is minus 4.0′ (be very careful to take this from the correct column because height of eye is given in both metres 'm' and feet 'ft').

(d)   After applying Dip look for the correction under SUN. Be sure you look at the correct column for the *month* in which you take the observation, also, whether you have taken the lower limb of the sun or the upper limb, because the first correction is plus, and the second is minus.

(e)   Having found the true altitude subtract it from 90° and obtain the zenith distance, not forgetting it is named the *opposite* to the bearing of the sun.

(f)   To the ZD apply the Dec., same names add, different names subtract, and the result will be the *latitude*; named N or S the same as the greater of the two.

A simple problem to work provided careful attention is given to detail and the rules about adding and subtracting is remembered.
Note. In some problems the DR latitude is not always given.

On May 4 1972 in DR Long. 2° 00′ E the sextant altitude of the sun's lower limb, when on the meridian, was 57° 10′ bearing south. Height of eye 18 feet, index error 1′ on the arc.

May 4 1972			
LMT Mer. Pass	1157	Sextant alt.	57° 10′ South
Long. in time E minus	−8	Index error	−1
GMT	1149	Observed alt.	57° 09′
Declination 1100	16° 03.9′N	Corr. for dip.	−4.1
'd' 0.7 correction plus	0.6	Apparent alt.	57° 04.9
Correct declination	16° 04.5′N	Correction plus	15.4
		True alt.	57° 20.3′S
			90°
		Zenith dist.	32° 39.7 N
		Declination	16° 04.5 N
		Latitude	48° 44.2′N

106

Notes on the working

(a) Mer. Pass. was 1157 and subtracting Long. in time (Long. east, Greenwich time least) gives GMT 1149.

(b) Dec. for 1100 GMT was 16° 03.9′ N and for the odd 49 minutes as little 'd' was 0.7 for the change in one hour, and 49 minutes equals 0.8 of an hour, 0.7 × 8 equals 0.56 (say 0.6), added as the Dec. is increasing.

(c) As the IE was 1′ on the arc, it is to be subtracted from the Sextant Angle. For the Dip correction, in the column headed DIP it will be seen that 18 ft. lies between 17.4 ft and 18.3 ft. the correction, therefore, is minus 4.1′.

(d) With an apparent altitude of 57 degrees 04.9′ look at the correction for the lower limb in the column headed April–Sept., and it will be seen that it lies between 57° 02′ and 61° 51′, giving a correction of plus 15.4′.

(e) With the true altitude subtracted from 90° obtain the ZD 32° 39.7′ N (named north, opposite to the sun's bearing south).

(f) To the ZD apply the Dec. (same names *add*) which gives the latitude.
Fig 85 illustrates the working of these two problems. S⊙ = Alt. Z⊙ = Zenith distance. Q⊙ = Decl. QZ = Lat.

If the same problem had been worked in the Southern Hemisphere with the sun bearing North of the observer the working would show as follows

Sextant alt.	57° 10′ North
Index error	−1
Observed alt.	57° 09′
Corr. for dip.	−4.1
Apparent alt.	57° 04.9
Correction plus	15.4
True alt.	57° 20.3 N
	90°
Zenith dist.	32° 39.7 S
Declination	16° 04.5 N
Latitude	16° 35.2 S

A simple problem sometimes given in the examination is to find the clock or chronometer time (GMT) of the sun's meridian passage. First, take the time of the Mer. Pass. from the *Nautical Almanac* for that day, and then apply the Long. in time. For example, if one was in Long. 12° W and the Mer. pass. was given as 1210, the GMT at the time of the sun's

107

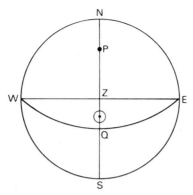

Fig 85 Finding latitude by meridian altitude of the sun NESW is the rational horizon; Z is observer's zenith, WQE is the equinoctial; WZE is the prime vertical; O is the Sun on the meridian; NZS is the observer's meridian; P the elevated pole; SO is the altitude of the sun; OZ is the zenith distance; OQ is the sun's declination; ZQ is the latitude of the observer; OP is the polar distance

Mer. pass. at that position would be 1210 plus 48 mins. (Long. 12° W in time) equals 1258 GMT.

Position Lines (Celestial and Terrestrial)

CELESTIAL

Having learned to find latitude by means of an observation of the sun when it is on the meridian, one must now learn how to find one's position by observations of the sun when it is out of the meridian.

The previous chapter explained how to use the ABC tables to find the true bearing of the sun (or azimuth) and a *position line* of the sun is drawn at *right angles* to the bearing (Fig. 86).

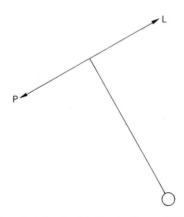

Fig 86 Shows the sun's true bearing in relation to the position line. O is the sun. PL is the position line; and the sun's true bearing is at right angles to that line

When taking the meridian altitude of the sun at noon to find Lat. the position line would run east and west, at right angles to the sun's true bearing, which is 180° when it is on the meridian south. Although this would give latitude, longitude would not be known unless another position line was available to cross with it, taken from an earlier observation (or later) and transferred as in a 'running fix.'

To work this type of problem we use what is called the *Intercept Method*. The formula is named after the navigator who devised it, Marc St Hilaire. The observed altitude is corrected to obtain the true altitude, which, subtracted from 90°, gives the zenith distance, which is known as the *true zenith distance (TZD)*. Using an assumed (or DR or estimated position) position, and by means of the Marc St Hilaire formula (see page 110) work out the calculated zenith distance (CZD). The difference between these two is the *intercept*.

1. The DR position is given for a problem.
2. With the GMT given find the correct Dec. and GHA.
3. Using the DR long. apply it to the GHA to find the LHA.
4. Correct the observed altitude to obtain the true altitude, and thence the true ZD(TZD).
5. With the LHA and Dec find the true azimuth from the ABC tables.
6. Using the Marc St Hilaire formula described below, find the calculated zenith distance (CZD) and thence the *intercept*.
7. To the DR position given in the problem apply the D. Lat. and the D. Long. using Traverse Tables with the bearing and intercept as the course and distance to find the position through which the position line is drawn. When the TZD and the CZD have been found learn that if the TZD is *greater* than the CZD the intercept is *away* from the Sun (A), and if the TZD is *smaller* (or tinier) than the CZD the intercept is *towards* the Sun (T), Think of TTT TRUE TINIER TOWARDS.

Study the drawing (Fig. 87) and it will be seen why this is so, and why it is important to name the bearing in reverse if the intercept is *away*.

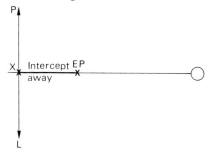

Fig 87   Use of the intercept method is illustrated (see text) O is the sun. EP the estimated position. EP—O the true bearing of the sun; EP—X is the intercept away. PXL is the position line

O   Sun
EP   Estimated position (DR position)
EP—O True bearing of sun
EP—X Intercept *away*
PXL   Position line

To work the problem of finding the calculated zenith distance (CZD) use the following mathematical formula. Learn the formula thoroughly which conforms to a set of rules based on the solution of spherical triangles.

Marc St Hilaire formula to find the CZD
What is required is the LHA, latitude and declination, then the value of the difference between the Lat. and Dec. (when of the same name the difference is found by subtraction (SS) and when of different names by addition).

> LHA   Log. haversine
> Lat.   Log. cosine
> Dec.   Log. cosine   _____

Total of sum equals Log. haversine

Change to	Natural haversine
(Difference) Lat. and Dec.	(Natural haversine_____
Add the two: equals	(Natural haversine_____of CZD

Now follow the working of the actual sum given the following factors: LHA 60° 20′ W DR Lat. 46° 52′ N Long. 41° 50′ W Dec. 10° 09′ N 62° 35′.

LHA 60° 20′ W	Log. hav. 9.40230	
Lat. 46° 52′ N	Log. cos. 9.83487	
Dec. 10° 09′ N	Log. cos. 9.99315	
	Log. hav. 9.23032	
Change to	Nat. hav. 0.16996	
(Lat. minus Dec.) 36° 43′	Nat. hav. 0.09920	Lat. and Dec. same names *subtract*
CZD	Nat. hav. 0.26916	
	CZD	62° 30.2′
	TZD	62° 35′
	Intercept	4.8′ *Away*

Using the tables A, B and C from *Norie's Tables* we calculate C to be .41 S. From table C at the back of this book will be found the true bearing to be S 74.4° W. South because C is named south and west because the LHA is west.

Having obtained the true bearing now lay this off through the DR Position 46° 52′ N, 41° 50′ W, and from this position *away* from the sun

110

measure off 4.8 miles, and through this position draw a position line at right angles to the true bearing. Normally, this would be laid off on the chart, but at the examination candidates may be required to lay it off on squared paper. Plot the DR position, and along the bearing (or the reverse if the intercept is away) of the sun, measure off the distance by counting the number of squares as miles. From this position count the number of squares vertically which will be equal to the D. Lat, and the number of squares horizontally will be the Dep. which must be changed into D. Long. by use of the Traverse Tables.

Below are some examples of plotting position lines by use of a chart (Fig 88) and on squared paper (Fig. 89 overleaf).

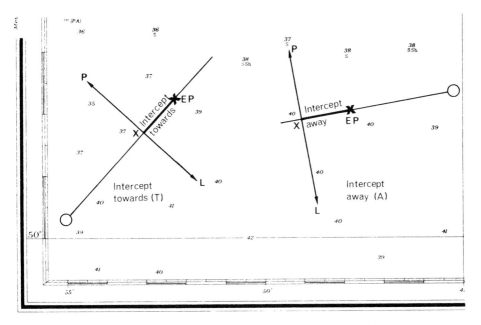

Fig 88 An example of plotting position lines by using a chart

The type of problem involving two separate observations of the sun with a run in between means that from the first estimated position (or DR position) lay off the intercept and through this position draw the first position line. From this lay off the course and distance steamed, at the end of which will be the second estimated position. The second bearing of the sun, with the intercept, is then plotted through this estimated position. The first position line is then transferred (as in a running fix) through the second estimated position. Where it cuts the second position line is the ship's position.

111

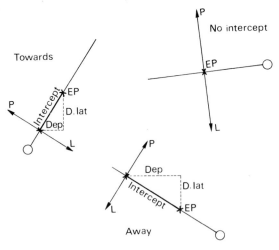

Fig 89 Positions plotted on squared paper

TERRESTRIAL POSITION LINES

In some problems a celestial position line is worked out as previously described, and a terrestrial position line is given by means of the bearing of a lighthouse, or a point of land, and the plotting of this on the chart is done in exactly the same way as a 'running fix', examples of which are next shown. In all these examples 'F' is the final position.

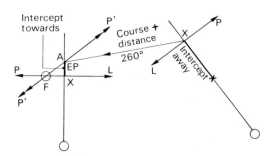

Fig 90 Plotting position lines

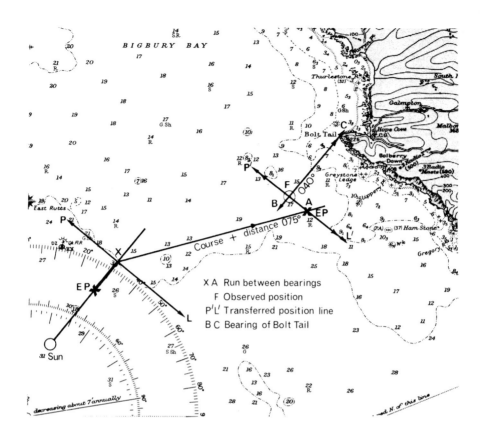

X A Run between bearings
 F Observed position
P'L' Transferred position line
B C Bearing of Bolt Tail

Fig 91 Terrestrial position lines crossed with a celestial position line

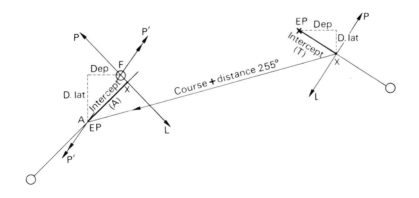

Fig 92 Celestial positions applied to squared paper

113

Chapter 8   Ship Stability

DEFINITIONS AND TERMS USED—UNDERSTANDING OF GENERAL
PRINCIPLES.

The stability of a ship depends upon the factors which tend to return it to
its normal position of flotation, whenever it is moved from that position
by forces such as wind and sea. With a floating object such as a ship the
volume of water it displaces is equal to the immersed part of the ship, and
the weight of the whole ship is equal to the weight of the amount of water
displaced.

However well a fishing boat may be designed and constructed, it is the
skipper's responsibility to ensure that she is always in a stable and sea-
worthy condition.

The chapter is designed to give a practical outlook concerning stability,
but calculations and theory must be left to larger and more sophisticated
text books on the subject. This advanced knowledge is normally only
required for higher fishing certificates.

Definitions

Centre of Buoyancy (B).   This is the geometrical centre of the underwater
part of a ship. The force of buoyancy acts vertically upwards through the
centre of buoyancy. When a ship heels, B moves to the low side in a line
parallel to the lines joining the centres of gravity of the emerged and
immersed wedge portions of the hull as illustrated below.

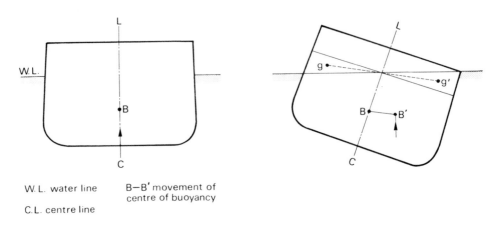

W.L. water line   B—B′ movement of
                      centre of buoyancy
C.L. centre line

Fig 93   Illustrating movement of centre of buoyancy

114

Centre of Gravity (G). This is the centre of the total weight of a ship and every item on board, and it is the point about which the ship would balance (see Fig 94). Also, it can be defined as the point through which all the weight of a ship is considered to act vertically downwards. G moves directly towards the centre of gravity of any weight added to a ship, and directly away from the centre of gravity of any weight removed from a ship. It is essential to learn this basic principle as it will help to understand the measures required in dealing with the stability of a fishing boat.

G also moves parallel to any weights moved on board, such as from the centre line to the sides of a boat.

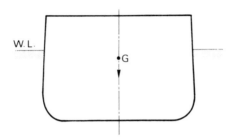

Fig 94 Centre of gravity of a vessel

Displacement. This is the actual weight of the ship and all its contents at any time, and as a floating body displaces its own weight of water it means that displacement is equal to the weight of water actually displaced by the ship (Fig 95).

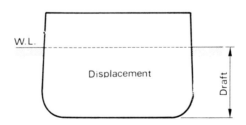

Fig 95 Illustrating displacement and draft

Draft. This is the depth of the bottom of the ship's keel below the surface of the water. Draft measurements must be read at the extreme ends of a fishing boat. If the draft is the same at each end the boat is on an even

5

keel but if one reading differs from the other the boat is 'trimmed' either by the head or the stern according to the readings. In reading the draft remember that the lower edge of the figure represents the reading in feet (or in the future metres).

Metacentre (M). The point at which a vertical line drawn upwards through the centre of buoyancy at a small angle of heel (up to about 15°) meets the ship's centre line is the metacentre.

Metacentric height (GM). This is the distance between the centre of gravity G and the metacentre M. It is called 'positive' when G is below M, and 'negative' if G is above M. If negative the boat would be in a state of unstable equilibrium (see Fig. 96).

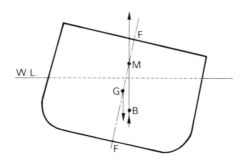

Fig 96 Showing metacentric height

Freeboard. This is the distance from the deck line to the water line. The continuous deck from which it is measured is called the freeboard deck.

Inertia. This is the resistance of a body to motion or to a change of direction. It requires a certain amount of force to overcome inertia as a stationary body will resist any attempt to move it, or if it is moving to change its speed or direction.

Initial Stability. This is the statical stability of a ship at a small angle of heel, and is indicated by GM (Metacentric height). If the GM is large the ship could be 'stiff' and if small the ship could be 'tender'. Normally it would not give any real indication as to how a fishing boat would behave at angles of heel beyond that used to find the Metacentre (M). A 'stiff' ship returns quickly to its upright position and creates an uncomfortable motion for those living on board, and also places a strain on gear and fittings such as the mast, *etc.* A 'tender' ship is slow to return to an upright position and if subjected to excessive outside

116

pressures, such as wind or seas, could develop a serious list, or even capsize. Without calculations to find the GM one can only 'feel' statical stability of a boat. Whether one considers a boat to be 'stiff' or 'tender' must depend upon experience. Each new fishing boat is supplied with stability details by the builders. These should be understood and one should find out what can or cannot be carried on board and where it should be stowed.

Range of Stability. This is defined as the angular range over which a boat will have positive statical stability. It indicates the angle to which a boat could heel before she would capsize. It is not possible to relate initial stability with the range of stability and the two cannot be connected. The range of stability can be increased considerably by an added amount of freeboard.

Reserve Buoyancy. This is the volume of a ship's hull, between the water line and the freeboard deck, illustrated in Fig 97.

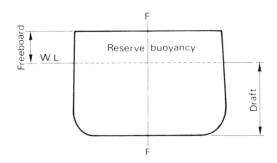

Fig 97   Showing reserve buoyancy

Righting lever. This is measured by the length of a horizontal line between the centre of gravity G and the vertical line through the centre of buoyancy B. It will be seen from Fig 98 how the amount of GM affects the righting lever GZ, or how the movement of G can increase or decrease the size of GZ (see Fig 98 overleaf).

Stable equilibrium. A fishing boat is said to be in stable equilibrium when, if heeled over by some external force, it would try to return to an upright position. In effect, it means that for small angles of heel she has a positive GM, and for any angle of heel the righting lever GZ is on the low side of the boat. Fig 99 shows varying angles A B and C affecting equilibrium.

117

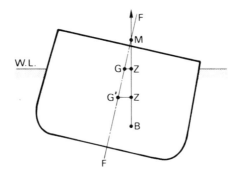

Fig 98   Righting lever

Neutral equilibrium. If a fishing boat, heeled by some external force, had no tendency to return to the upright or to heel still further, she is said to be in neutral equilibrium. This would occur when G and M coincide. (B).

Unstable equilibrium. A fishing boat would be in a state of unstable equilibrium when, if inclined by some external force, she would heel over still further, and may possibly capsize. For small angles of heel a boat would have to have a negative GM, and the righting lever GZ would be on the high side (see sketch C). For larger angles of heel this is not necessarily so as B could move further outward and eventually become vertically under G. At the larger angle of heel a boat may once again become stable but would be in the condition known as 'loll', that is she would oscillate (swing like a pendulum) about this new position when affected by external forces.

Free surface liquids. If a compartment or tank on a fishing boat is filled completely with water or oil the effect upon stability is the same as that of a solid, and would have the same effect upon the centre of gravity G as any other weight added. However, if the compartment or tank is only partially filled with liquid the surface of that liquid is free to move when the boat heels over. A partially filled tank is described as being 'slack' and if the boat is heeled the liquid will move to the low side. Free surface water or oil can build up in tanks, fishrooms, fish pounds, bilges, accommodation, and on deck when scuppers and wash ports have become blocked. The effect of free surface liquid is virtually to move the centre of gravity of a boat upwards to a dangerous extent. The centre of gravity of the liquid is transferred to the low side of the boat G^1 which reduces the righting lever GZ and has the same effect as if G had moved up to G^2 in the figure below.

118

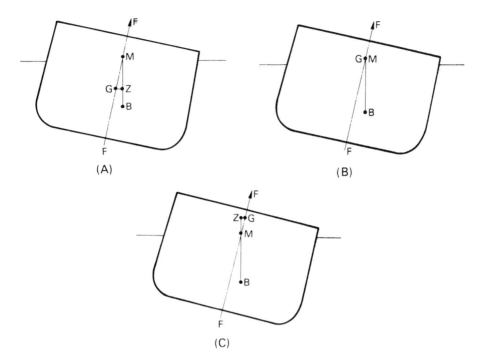

Fig 99 Showing stable equilibrium in (A) neutral equilibrium in (B) and unstable equilibrium in (C) of diagram

As stated earlier if G moves upwards it reduces the GM and thence initial stability. The effect of free surface liquids does not depend so much upon the quantity of liquid in the tank or compartment as upon the actual area of the free surface liquid, or the waterplane as it is called. The free surface effect of a few inches of liquid could be the same as if there were several feet of liquid in the same tank or compartment (see Fig 100).

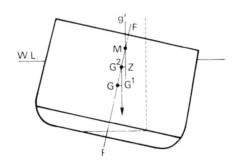

Fig 100 Effect of free surface liquids is shown here

119

The smaller the tank or area the less effect free surface liquids will have upon stability, and by dividing a tank in half with a fore and aft bulkhead it will reduce the effect to one quarter, and if the tank is sub-divided by two fore and aft bulkheads it will reduce the effect to one ninth, as compared with that of an open undivided tank or compartment.

Tanks should be either empty or full, but obviously, this is not always practicable, and the fitting of one or more fore and aft bulkheads will always reduce the effect of free surface liquids.

In stability calculations the height of the metacentre (*not* the metacentric height which is indicated by GM) and the centre of gravity is indicated by their height above the keel K. KM and KG are used to indicate these heights.

It will have been seen from the diagrams in this chapter that in a stable ship the metacentre is above the centre of gravity, and the centre of buoyancy is below it. Remember MGB which can stand for

*M*otor
*G*un
*B*oat   in this order.

The initial GM in a boat's light condition will be given in the plans and details handed over by the builders when a boat is completed. M will remain constant but G can move up or down or sideways according to the distribution of weights around the boat. While the GM varies according to the design of each fishing boat, about 1.5–2.0 feet can be taken as the normal when a boat is in a stable condition.

If too much weight is placed high up in a fishing boat G will move upwards towards M and GM will be reduced. If weight is removed from below G it will again move upwards. A number of fishing boats have been lost in the past because of a lack of knowledge of what happens when weights are added or removed on board, and the effect of the shift of G.

To summarise the causes of instability:

1 Weights may be placed too high in a fishing boat when fish or shell-fish is being handled, and eventually cause a negative GM to appear.
2 Fuel, water, oil and stores used at sea may be taken from positions low down in the ship below G, causing G to rise with the result a negative GM is formed.
3 Unexpected weights may be added high up, such as ice accretion on the masts, rigging, bridge and upper works.
4 The effect of free surface liquids in slack tanks, or compartments, or on deck.

Instability can lead to a negative GM and the cure for this always must be to lower the centre of gravity, but in doing this care must be taken not to make a dangerous situation even worse. Adding weight as low as possible and removing it from as high as possible should be the guiding rule.

Taking the following actions can assist in eliminating or rectifying the above causes of instability:

1 Filling a tank is one practical method (some larger fishing boats are fitted with double bottom tanks for the carriage of water ballast or oil). If this tank is subdivided at the centre line commence filling the low side first, and when this is about two thirds full commence filling the high side. The boat might list slightly more at the beginning but the overall effect will be to lower G, whereas if the tank was filled the high side first, the boat might heel over suddenly and violently from one side to the other, with the possibility of capsizing.
2 Never pump out a lower tank or double bottom tank to correct a boat heeling because this could cause G to rise still further.
3 Any fish, or shellfish stowed on deck should be put below into the fish holds as quickly as possible, or if the situation is urgent and dangerous it should be jettisoned.
4 Remove ice accretions, always starting at the highest point.
5 Clear all free surface water from accommodation, decks, fish pounds, fish rooms and bilges.
6 All loose gear or drums of lubricating oil, stores, *etc.*, should be stowed below decks, and as low down as possible.

Fig 101 illustrates some of the measures to be taken (see next page).

If a fishing boat is initially stable and develops a 'List' because of the distribution of weights on board this can be corrected in three different ways:

1 Move weight from the low side towards the high side.
2 Add the equivalent weight on the high side.
3 Remove the weight from the low side (or from the off centre position).

If, on the other hand, a fishing boat is initially unstable and heels over this must be corrected by entirely different means. A boat is said to be in a state of 'loll' when this happens but some examiners dislike the use of this term and prefer 'in a state of unstable equilibrium', a definition given previously.

1 Move weight from high up to low down in the boat.
2 Add additional weight low down in the boat.
3 Remove weights which are high (*eg* ice on rigging, fish or shellfish piled high on deck).
4 Reduce the amount of free surface water.

121

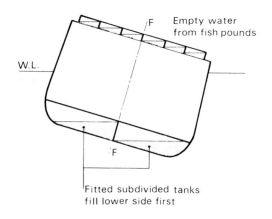

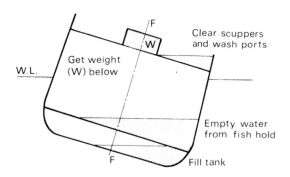

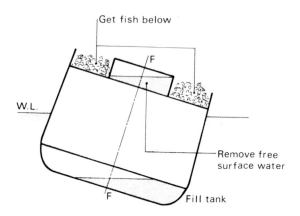

Fig 101 Measures to be taken to secure stability (a) top, (b) centre and (c) bottom, (see text on previous page)

122

Finally, consider some examples of how the stability of a fishing boat can be affected adversely:

1 Heavy seas entering the accommodation or engine room, penetrating into stores or fish holds and creating a free surface effect.

3 Water accumulating on deck due to blocked scuppers and wash-ports.

3 Ice accumulating on the upper works and rigging, leading to G being raised sufficiently to give a negative GM.

4 Gear becoming fast upon an obstruction. The pull of the warps at the gallows head (if it had not been possible to reverse the winch or cut the warps), producing an off centre loading effect, and possibily listing the boat towards heavy seas which might come aboard in great force and create the situation described in No. 1.

5 Overloading fish or shellfish on deck which could slide quickly to the low side if the boat heels over, thus creating almost the same effect as free surface water.

When weights are suspended at a height such as when the cod end is hoisted inboard, the effect is the same as if the weight had been transferred *to that height*, and G would rise accordingly. It can be a crucial moment when the full weight of the cod end is suspended from a block high up the mast, and it must be appreciated that the stability of a boat is considerably lessened at that moment (Fig 102).

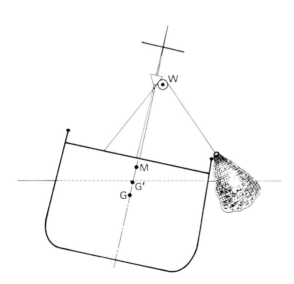

Fig 102 Effect of hoisting trawl on stability — W shows weight effect, G weight moves to G' and reduces GM

Chapter 9   Miscellaneous

USE AND READING OF THE ANEROID BAROMETER — UNIFORM
SYSTEM OF BUOYAGE AND WRECK MARKING — NAVIGATIONAL
MARKS.

Use and Reading of the Aneroid Barometer

An ability to read the aneroid barometer should be allied to an understanding of its practical use and some basic knowledge of meteorology.

The aneroid barometer is used for measuring atmospheric pressure, and
depends upon the slight movements of the top of a disc-shaped metal
box. This is made of a very thin corrugated metal from which air is
partially exhausted, and changes in the atmospheric pressure causes the
top of this metal box to move. By a system of springs and levers this
movement is transmitted to a pointer on the face of the barometer. The
atmospheric pressure existing at that time and in that place can be read
from the graduated dial marked on the face of the barometer.

A typical aneroid barometer is illustrated in Fig 99. It will be seen that
the face is graduated in millibars, some barometers being graduated in
both millibars and inches.

Fig 103 The Aneroid barometer

The accuracy of a boat's aneroid barometer can be checked by contacting the local Port Meteorological Officer, situated at the principal sea ports, or with other authorities such as the local Harbour Master, Coastguards, etc.

The only correction needed is for height. The higher one is above sea level the less is the density of the air, and the barometer would fall. The correction amounts to 0.37 millibars for every 10 feet of height, which must be added to the barometer reading to obtain the correct reading at sea level.

For example, if an aneroid barometer situated in a Coastguard Station 200 feet above sea level was required to be corrected for a sea level reading, it would mean 20×0.37 equals 7.4 millibars. This amount would have to be added to the readings as in Fig 104.

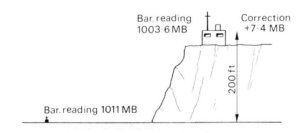

Fig 104 Checking barometer with coastguard station above sea level

Normally, in fishing boats the aneroid barometer can be accepted as being at sea level and no correction is necessary.

If, when comparing a barometer reading with an official reading given by one of the previously mentioned authorities, a difference is found, the pointer on the face of your barometer can be adjusted as follows. At the back a small hole will be found in which is situated a screwhead. By gently turning this screwhead with the aid of a small screwdriver the pointer can be moved to the correct reading, but turn the screw only by a very small amount.

The normal atmospheric pressure in waters around Great Britain is shown as a barometer reading of 1014 millibars, or 29.94 inches, so that any readings above or below these figures indicate a change of atmospheric pressure from the normal.

The pressure in certain places in the World is semi-permanently above the mean, while in other places it is semi-permanently below the mean. These places are referred to as regions of high and low pressure respectively. There are also temporary areas of high or low pressure.

Because of the rotation of the Earth, air which is drawn towards a centre of low pressure is deflected to the right in the northern hemisphere, and to the left in the southern hemisphere. The result is an anti-clockwise circulation of wind around and area of low pressure in the northern hemisphere and a clockwise circulation in the southern hemisphere. Circulation around areas of low pressures are termed cyclonic. Conversely the wind circulates in a clockwise direction around an area of high pressure in the northern hemisphere and in an anti-clockwise direction in the southern hemispheres, such circulation being term anticyclonic.

By reading an aneroid barometer at regular intervals (more frequently in bad weather) it is possible to understand certain basic principles of what the weather is likely to do.

1 Low pressure shows unstable and changing conditions.
2 High pressure shows stable and continuing good conditions.
3 Steady rise shows good weather approaching.
4 Steady fall shows bad weather approaching.
5 Rapid rise shows better weather may be only temporary.
6 Rapid fall shows stormy weather approaching quickly.

It is important to have an idea of what weather to expect on passage to or from fishing grounds and when actually fishing.

Try to avoid being caught out by bad weather at sea, but if it is unavoidable then a knowledge of how the weather is likely to act is essential to decide whether to ride it out or try to make for shelter. One's own local weather knowledge will help, aided by the official weather forecast through the radio broadcasts and gale warnings, and in conjunction with barometer readings. A study of the sky is also required and a careful watch on the direction of the shift of the wind. Bear in mind that single weather observer forecasting cannot expect too much success beyond a period of about 6 hours, so that regular listening to weather forecasts is essential.

Surface friction has two effects on the wind. Firstly, it causes a reduction in the strength of the wind at the surface, and secondly it causes the wind to be deflected some 10° to 20° across the isobars, inwards towards the centre of low pressure, or outwards away from the centre of high pressure. Buys Ballot's Law sums this up as follows: 'If you face the wind the centre of low pressure will be approximately 90° on your right hand in the northern hemisphere, and 90° on your left hand in the southern hemispheres. See Fig 109 showing a depression in the northern hemisphere.

Fig 105 shows how Buys Ballot's law operates. In a depression the wind revolves anti-clockwise around the centre, so that in whatever position one is situated relative to the centre the direction of the centre remains the same. In the southern hemisphere the reverse is the case.

Depressions or low pressure areas always move in certain directions,

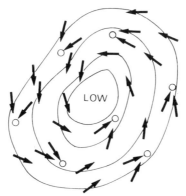

Fig 105   Illustrating Buys Ballot's law for finding the centre of a depression
(Observer arrowed O)

and it is important to know whether a depression is going to pass over or to the north or south of one. An understanding of Buys Ballot's law helps to determine this, apart from the knowledge of future wind directions received from weather forecasts.

It must be appreciated that weather forecasting is not an exact science despite the high standards reached, because weather changes can be unpredictable, as due to unknown factors a depression can change its speed and direction suddenly and unexpectedly.

Generally speaking depressions approach Great Britain from the Atlantic travelling in an easterly direction, but on occasions depressions have moved in quite unforeseen directions, such as westerly, due north, or south, *etc.* This is where sudden changes of wind, allied to barometer readings, can be of assistance in determining the possible change in weather conditions (see Fig 106).

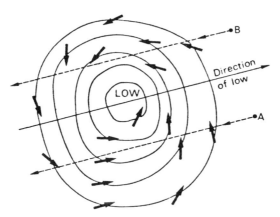

Fig 106 Diagram illustrates wind direction in surroundings of a depression (see text)
(Northern hemisphere)

127

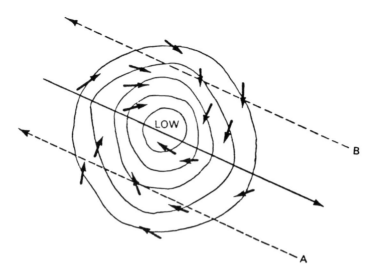

Fig 107 Diagram illustrating wind direction in surrounding areas of a depression (Southern hemisphere)

From Fig 106 it will be seen that a fishing boat situated at A. can expect the wind to veer from south east through south to south west and west. Whereas a fishing boat situated at B. would expect the wind to back from south easterly through north east and north to north west.

Isobars are lines of equal barometric pressure and join up all those positions which have the same atmospheric pressure and are shown on weather charts. The closer the isobars are drawn together the greater the rise or fall in the atmospheric pressure in that area, and therefore, the stronger the winds (see Fig 108).

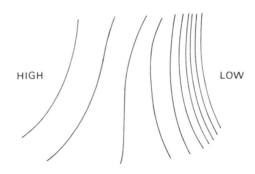

HIGH   LOW

Fig 108   Inclination of isobars — high and low

Wind speed is indicated by the Beaufort Scale. This should be memorised in order to understand the details given out in the weather forecasts.

Beaufort number	Windspeed in knots	Descriptive terms
0	Less than 1	Calm
1	1–3	Light air
2	4–6	Light breeze
3	7–10	Gentle breeze
4	11–16	Moderate breeze
5	17–21	Fresh breeze
6	22–27	Strong breeze
7	28–33	Near gale
8	34–40	Gale
9	41–47	Strong gale
10	48–55	Storm
11	56–63	Violent storm
12	64 plus	Hurricane

When determining the wind direction and speed remember to eliminate the effect of speed of the boat. For example, if heading into a 20 knot wind and the speed was 10 knots the effect would be 20 plus 10 and the effect of a 30 knot wind would be felt on board or, in other words, a near gale; whereas if running before a wind of this strength the effect would be 20–10 equals 10 knots, or a gentle breeze.

The opposite to a low pressure area is a high pressure area and the wind revolves round a high pressure area in a *clock-wise* direction in the northern hemisphere as shown and in the southern hemisphere the winds will revolve *anticlockwise* (see Fig 109).

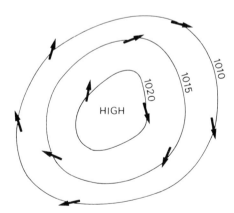

Fig 109   High pressure winds revolve clockwise in the northern hemisphere as shown and in the southern hemisphere the winds will revolve anticlockwise

When using the terms 'veering' and 'backing' understand that the wind is named from the direction it blows. When reference is made to a south west wind it means that it is blowing from the south west in a north east direction. A wind is said to be 'veering' when it changes direction from SE to S and SW, or W to NW. 'Backing' means the wind changing from SW to S, or NE to N, *etc*.

An understanding of a few of the most important technical terms used in meteorology will assist in understanding weather forecasts. These are:

1 *A cold front*. The line between the advancing cold air at the rear of a depression and the warm frontal sector.
2 *Depression*. An area associated in the barometric pressure field by a system of close isobars, having the lowest pressure on the inside of the area.
3 *Front*. The line of separation at or near the earth's surface between warm and cold air masses.
4 *Further outlook*. Generally added to a detailed forecast and giving conditions likely to be experienced within the next 24 hours or more, following the period covered by the actual forecast.
5 *Secondary depression or 'secondary'*. The isobars around a depression are not always even and sometimes they show bulges or distortions and may enclose separate centres of low pressure areas, and have quite separate wind circulations (Fig 110).

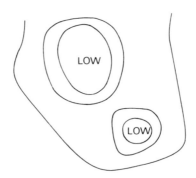

Fig 110 Illustrating a secondary low

6 *Squall*. A very strong wind which increases (and decreases) suddenly. Usually, it indicates a change of wind direction.
7 *Tendency of the barometer*. The amount of change in barometric pressure in the 3 hours preceding the time of the observation. It is given in such terms as 'rising', 'falling', 'steady', *etc*.

130

Depression

A more detailed explanation of depressions can be obtained by a study of the section 'Through a depression', in both the Northern and Southern Hemispheres.

The 'warm front' is indicated by the symbols ● ● ● ● and the 'cold front' by ▲ ▲ ▲ ▲ (think of icicles). In the section through the depression the type of weather to be expected in the warm front and the cold front is shown.

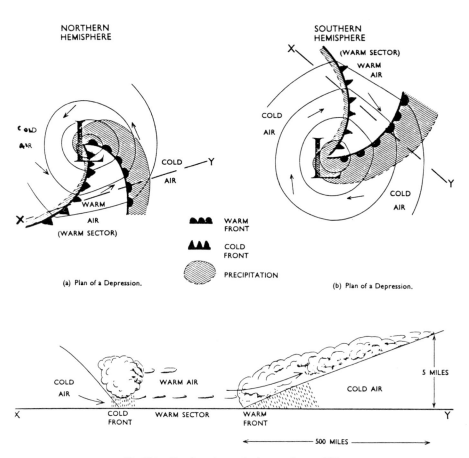

Fig 111   Section through depression at XY

Occlusion

In some circumstances the cold front moves faster than the warm front and gradually overtakes it, causing the warm air to be lifted up from the surface. When this happens the depression is said to be 'occluded' and the front have merged into an 'occlusion'.

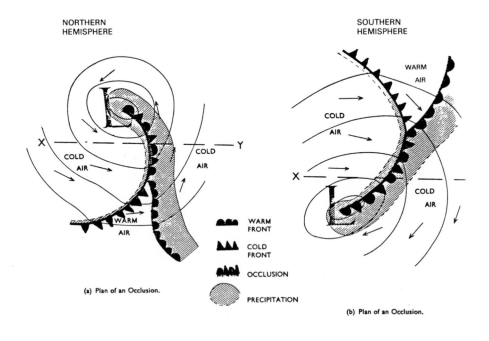

NORTHERN HEMISPHERE

SOUTHERN HEMISPHERE

WARM AIR

COLD AIR

COLD AIR

COLD AIR

COLD AIR

WARM AIR

COLD AIR

WARM FRONT

COLD FRONT

OCCLUSION

PRECIPITATION

(a) Plan of an Occlusion.

(b) Plan of an Occlusion.

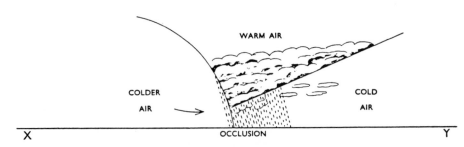

WARM AIR

COLDER AIR

COLD AIR

X   OCCLUSION   Y

Fig 112   Section through occlusion at XY

Precipitation

This is another term for rain, dew, *etc.*

Clouds

A knowledge of the types of clouds can give a much better indication of possible changes in weather. There are publications available which give photographs of the various cloud formations but they can be summarised as follows:

132

Upper clouds (above 18,000 feet), cirrus (Ci.), cirro–cumulus (Ci.–Cu.) and cirro–stratus (Ci.–St.) middle clouds (8 000 to 18,000 feet), alto–cumulus (A.–Cu.), alto–stratus (A.–St.), nimbo–stratus (N.–St.), low cloud (below 8,000 feet), cumulo–nimbus (Cu.–Nb.), stratus (St.) cumulus (Cu.) strato–cumulus (St.–Cu.), nimbus (Nb.) fracto–nimbus, often called 'scud'.

Two major dangers facing fishing boats are ice and fog.

In certain weather conditions ice accumulating on the hull, super-structure, and rigging of a fishing boat can be a serious danger to stability The accumulation can occur from three causes:

1 Fog, with freezing conditions, including frost 'smoke.'
2 Freezing drizzle or freezing rain.
3 Sea spray or sea water breaking over a ship when the air temperature is below the freezing point of sea water (about minus 2° C).

The weight of ice accumulating in rough weather with very low tempera-ture and large amounts of spray and heavy seas breaking over a ship can be rapid. The speed with which ice accumulates increases progressively as the force of the wind increases and as the sea temperature decreases. When these conditions occur speed must be reduced and, if necessary, one should heave to so as to avoid heavy spray and seas breaking on board. Alternatively, seek shelter or steer towards warmer water if the direction of the wind and sea will allow this.

Fog is caused at sea by changes in the air temperature relative to the temperature of the sea. It is described in technical terms as the cooling of the air in contact with the surface of the sea to a temperature at which it can no longer maintain, in an invisible state, the water vapour which is present in it. Condensation of this vapour into minute droplets produces fog. If warm, moist air, flows over a cold sea surface it will produce what is called 'sea fog', or 'advection fog'. Again, if very cold air flows over a much warmer sea surface a fog is produced which gives the appearance of the sea steaming.

Warnings are obtained from weather forecasts about conditions likely to cause fog, but one must observe rapid changes in temperature and visibility.

IALA Maritime buoyage System 'A'

Implementation of the new system 'A', which is a combination of the Cardinal and Lateral Systems, see L70 shown on page 138, began in April 1977 and will take about four years to complete. Work began in the eastern part of the English Channel area, and southern part of the North Sea, eventually extend-ing to all coasts of the British Isles, the North Sea, and the Baltic.

IALA Buoyage System A

LATERAL MARKS

Used generally to mark the sides of well defined navigable channels

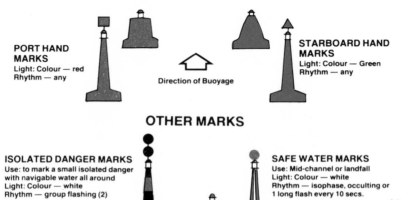

PORT HAND MARKS
Light: Colour — red
Rhythm — any

Direction of Buoyage

STARBOARD HAND MARKS
Light: Colour — Green
Rhythm — any

OTHER MARKS

ISOLATED DANGER MARKS
Use: to mark a small isolated danger
with navigable water all around
Light: Colour — white
Rhythm — group flashing (2)

SAFE WATER MARKS
Use: Mid-channel or landfall
Light: Colour — white
Rhythm — isophase, occulting or
1 long flash every 10 secs.

SPECIAL MARKS   Any shape permissible
Use: of no navigational significance
Light: Colour — yellow
Rhythm — different from any white lights used on buoys

CARDINAL MARKS

Used to indicate the direction from the mark in which the best navigable water lies, or to draw attention
to a bend, junction or fork in a channel, or to mark the end of a shoal.

LIGHTS: ALWAYS WHITE

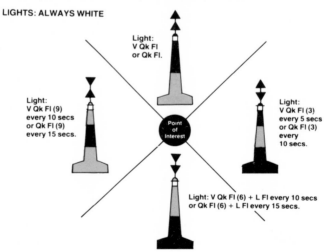

Light:
V Qk Fl
or Qk Fl.

Light:
V Qk Fl (9)
every 10 secs
or Qk Fl (9)
every 15 secs.

Point
of
Interest

Light:
V Qk Fl (3)
every 5 secs
or Qk Fl (3)
every
10 secs.

Light: V Qk Fl (6) + L Fl every 10 secs
or Qk Fl (6) + L Fl every 15 secs.

Fig 113   I.A.L.A. buoys

Buoys and Beacons   L70
IALA Buoyage System 'A'

The combined Cardinal and Lateral System (Red to Port)

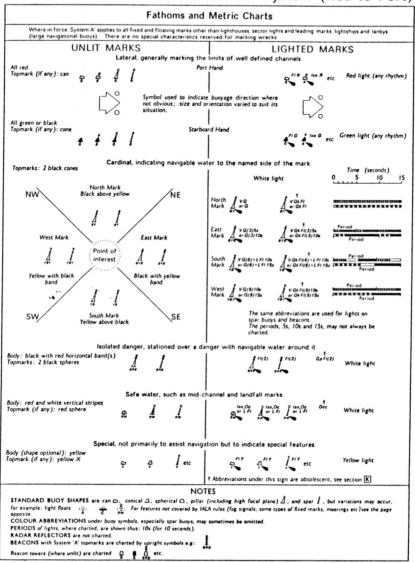

Fig 114   I.A.L.A. buoys and beacons on charts. Reproduced from B.A. publication No. 5011 with the permission of the Controller, H.M. Stationery Office and the Hydrographer of the Navy

7

135

Admiralty hydrographical publication *NP 735* contains full details of the new system.

System 'A' applies to all fixed and floating marks, other than lighthouses, sector lights, leading lights and marks, and lightships, and will indicate the sides and centre lines of navigable channels, natural dangers, and other obstructions such as wrecks, *etc.*

The system provides five types of marks which may be used in any combination.

Lateral marks incidate the port and starboard hand sides of channels.

Cardinal marks will indicate the navigable water which lies to the named side of the mark used in conjunction with the compass.

Isolated Danger marks will be erected on, or moored on or over dangers of a limited extent,

Safe Water marks such as mid-channel buoys, or as a land fall buoy, indicating safe water all round.

Special Marks will indicate special areas which will be clearly indicated on charts and will include spoil ground marks, traffic separation marks, military exercise areas, cable, pipeline or outfall pipes, and recreation zones areas. The characteristics of these marks will depend upon colour, shape, and topmarks by day, and light colour and rhythm by night.

The five basic buoy shapes are can, conical, spherical, pillar, and spar, with colours red and green for the port and starboard hand *Lateral Marks,* and yellow for the *Special Marks.*

Other types will have horizontal bands or vertical stripes as shown on the following pages.

For top marks can, conical, spherical, X shaped will be used.

The colour of lights, where used, will be red for port and green for starboard on *Lateral Marks,* and yellow for *Special Marks.* Other types will have a white light, distinguished by its rhythm.

A *port hand* mark is coloured *red* and its basic shape is *can,* for either buoy body or topmark, or both.

A *starboard hand* mark is normally coloured *green* and its basic shape is *conical,* for either buoy body or topmark (point up), or both.

By night a port hand buoy is identifiable by its red light and a starboard hand buoy by its green light; any rhythm (flashing, occulting, isophase, *etc.*) may be used.

The lateral colours of red or green will frequently be used for minor shore lights, such as those marking pierheads and the extremities of jetties. In British waters, to avoid confusion with the navigation lights of ships, minor lights, if fixed, will be shown in pairs, disposed vertically. Alternatively, single red or green lights will be flashing of occulting.

Throughout the many areas in which System 'A' is being introduced there are variations in the design of equipment in use. It is emphasised, therefore, that the illustrative diagrams are in general conformity with the approved shapes, colouring and topmarks, but are not intended to convey the detailed configuration, the exact colour disposition and the topmark size of the buoys in use; these features will vary somewhat, depending on the individual design of the buoys in use.

It should be noted that in particular instances starboard hand buoys may be exceptionally coloured black instead of green, but not in the United Kingdom.

The local direction of buoyage will be taken as approaching a harbour, river estuary, or other waterway from seaward. The general direction of the buoyage system around the British Isles will run northward along the west coast and through the Irish Sea, eastward through the English Channel, and northward through the North Sea.

If, in some places a strait is open at both ends the local direction of the buoyage may be over-ridden by the general direction.

If any doubt exists as to the direction of the buoyage system the following mark will be shown on the chart.

Fig 115   Symbol to indicate buoyage direction

In future, on charts, the light star symbol inserted above buoy symbols will be discontinued, and this will enable top mark symbols to stand out more clearly. The magenta light flares will now be inserted with their points adjacent to the position circles at the base of the buoy, and this will avoid obscuring the top marks symbols.

Fig 116   Buoyage symbols

The colours of buoys are indiciated by the following abbreviations: W. white, B. black, R. red, Y. yellow or amber, G. green, Gy. grey, Bu. blue.

The following illustrations of the System 'A' buoys show clearly the types, colours, top marks, and light characteristics.

New Dangers

Definition. A newly discovered hazard to navigation not yet shown on charts, or included in sailing directions, or sufficiently promulgated by notices to mariners, is termed a New Danger. The term covers naturally occurring obstructions such as sandbanks or rocks, or man-made dangers such as wrecks.

Marking. A New Danger is marked by one or more Cardinal or Lateral marks in accordance with the System 'A' rules. If the danger is especially grave, at least one of the marks will be duplicated as soon as practicable by an identical mark until the danger has been sufficiently promulgated.

Lights. If a lighted mark is used for a New Danger, it must exhibit a quick flashing or very quick flashing light: if it is a Cardinal mark, it must exhibit a white light; if a Lateral mark, a red or green light.

Racons. The duplicate mark may carry a racon, coded W ($\cdot--$), showing a signal length of one nautical mile on a radar display.

Fig 117   New dangers — Buoyage

A special mark may be used to indicate to the mariner a special area or feature, the nature of which is apparent from reference to a chart, sailing directions or notices to mariners.

Uses include:

Ocean Data Acquisition Systems (ODAS), *ie* buoys carrying oceanographic or meterological sensors;
Traffic separation marks;
Spoil ground marks;
Military exercise zone marks;
Cable or pipeline marks, including outfall pipes;
Recreation zone marks.

Another function of a Special mark is to define a channel within a channel. For example, a channel for deep draught vessels in a wide estuary, where the limits of the channel for normal navigation are marked by red and green Lateral buoys, may have the boundaries of the deep channel indicated by yellow buoys of the appropriate Lateral shapes, or its centreline marked by yellow spherical buoys.

Navigational Marks

Full details of all navigational lights are given in the *Admiralty List of Lights*, and in *Chart 5011 (Symbols and Abbreviations used on Admiralty Charts)* a full description is given of details of the various types of lights shown from light-houses, light vessels, and beacons.

The position of a lighthouse or light beacon is indicated on the chart by a star * and is further emphasised on the chart by overprinting with a magenta splash mark.

A light is described in the following terms:
1. Its distinguishing appearance or characteristic.
2. Colour.
3. Time required for one complete cycle (or period).
4. Elevation.
5. Range

The characteristics used to describe lights are as follows, the abbreviation being given first: (turn also to the heading '*Lights*' for a fuller description).

F	Fixed.
Oc.	Occulting.
Oc. (2)	Group Occulting.
Fl.	Flashing.
LFl.	Long Flashing.
Iso.	Isophase.
Q.	Quick Flashing.
VQ.	Very Quick Flashing.
UQ.	Ultra Quick Flashing.
IQ.	Interrupted Quick Flashing.
Alt.	Alternating.
Al.WR	Group Flashing.
F.Fl.	Fixed and Flashing.
Mo. (A)	Morse code light (with flashes grouped as in letter A).
Dir.	A directional light showing a narrow sector in one direction only as a single leading light.

Lights

These are:

Fixed	A continuous steady light (white or coloured).
Occulting	A steady light with at regular intervals, a sudden and total eclipse, the duration of darkness always being *less* than the duration of light.
Flashing	A light showing a single flash (white or coloured) at regular intervals, the duration of light being *less* than that of darkness.
Isophase	A light with a duration of light and darkness *equal*.
Quick flashing	Flashing continuously at more than 60 times a minute.
Interrupted quick flashing	Flashing at a rate of more than 60 times a minute with, at regular intervals, a total eclipse.

Alternating	A light which alters in colour, in successive flashes or eclipses.
Group occulting	At regular intervals two or more sudden eclipses in a group.
Group flashing	At regular intervals two or more brilliant flashes in a group.
Ultra quick flashing	Flashing continuously at 160 or more times a minute.
Long-flashing	Flash of 2 seconds or longer.
Fixed and flashing	A steady light with, at regular intervals, one flash of increased brilliance.
Fixed and group flashing	A steady light with two or more brilliant flashes in a group.
A morse code light	The characteristics of these lights are shown by the appropriate letter or figure in brackets, *eg.* MO (A).

The *period* of a light is the time shown against the light, *eg* Gp. Fl.(2) ev. 10 sec. indicates that the period of that light is 10 sec. and is the time occupied in the exhibition of the whole of the changes (flashes and darkness added together). For a light showing a single flash it would be from the *commencement* of one flash to the *commencement* of the next, *eg* Fl. ev. 15 sec. (a period of 15 sec.).

Fog Signals

Fog signals made by buoys, light vessels, lighthouses, *etc,* are defined as follows:

Explos.	Explosive fog signal. Signals explode in mid-air.
Dia.	Diaphone. Powerful low note terminating in a kind of grunt.
Siren	Fog siren. As medium powered high or low note, or a combination of both.
Horn	Fog horn. Powerful medium pitched note.
Bell	Fog bell. Varying note and power. May be operated mechanically or by wave action in which case operates irregularly.
Whis.	Fog whistle. Lower power and low note.
Reed	Reed. Low powered high 'piping' note.
Gong	Fog gong. As for bell.
Mo.	Morse code fog signal. Sounds one or more morse code letters or figures.

The colour of lights is white unless otherwise stated.

Elevation. This of a light is measured in feet (or metres) between the focal plane of the light and the level of mean high water springs.

Range. The range of lights is given in sea miles calculated as seen from a height of 15 feet above sea level (5 metres in the case of countries using the metric system). Two ranges are now quoted in official publications—1. The geographical range which is computed from the height of the light, the height of the observer, and the curvature of the earth; 2. The luminous range which is computed from the intensity or candle power of the light. Another term given to this is nominal range, which is the same as the luminous range when the visibility is 10 sea miles. These ranges will be changed gradually as information is received to conform to the new system of showing the nominal (luminous) range of the new metric charts and in the *Admiralty Light Lists*.

Sector Lights. These show various colours or characteristics in different directions, and the arc of each portion of the 360° circle that shows such a colour is called a Sector (see Fig. 118).

Leading Lights. These may be of any colour, and either fixed, flashing or occulting. Their object is to indicate a safe course into a harbour or through a channel. The rear light is always higher than the front light (see Fig. 118).

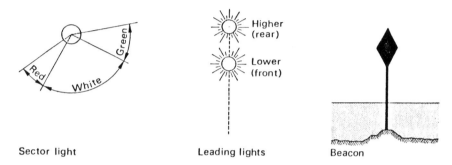

Sector light   Leading lights   Beacon

Fig 118   Leading lights and sector lights

Unwatched Lights. These are shown as (U) and indicate that no special watch is kept on this light and, therefore, cannot be relied upon implicitly. This term is now obsolescent and is being omitted from all future charts.

All bearings of lights are given as true from seaward (*ie* from the ship), and in the 0–360° notation.

Lightvessels

A lightvessel is indicated on the chart by the symbol shown in Fig 119, and a small circle in the base indicates its exact position.

A white riding light is exhibited from the fore-stay of each lightvessel at a height of 6 feet above the rail which shows in which direction the vessel is riding. If, from any cause a light vessel is unable to exhibit her usual characteristic lights whilst at her station, the riding light only will be shown. However, when a lightvessel is driven from her proper station to one where she is of no use as a guide to shipping she will display the following signals. The characteristic light will not be exhibited and a fixed red light will be exhibited at each end of the vessel. Red and white flares will be shown simultaneously every 15 minutes or at more frequent intervals on the near approach of any vessel.

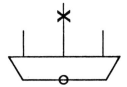

Fig 119   Symbol of a lightvessel

Fig 120   Lightvessel with watch buoy

Daymarks of Lightvessels. Lightvessels are painted red with their name on each side in large white letters. When a lightvessel is driven from her proper station by day two large black balls will be exhibited, one forward and one aft, and the international code flags LO, indicating 'a lightvessel out of position.' will be hoisted where they can best be seen.

Fog Signals of Lightvessels. All lightvessels exhibit their lights in foggy weather, in conditions of poor visibility and sound their fog signals. If, during fog or low visibility, a vessel is heard approaching too close to a lightvessel so as to involve risk of collision a bell will be rung rapidly during the silent intervals between successive soundings of the fog signal until all risk of collision is past.

Danger Signals from Lightvessels. A lightvessel will give warning to a ship running into danger by hoisting the flag U of the international code 'you are running into danger'. In addition, a gun or rocket may be fired or a sound signal given at short intervals, the letter U in morse or given by light or sound. The signal PS 'you should not come any closer' may be used also if a vessel is seen to be coming dangerously close to the lightship.

Lanby Buoys

A Lanby buoy (Large Automatic Navigation Buoy) is about 80 tons in weight and 40 feet in diameter, with a built up super-structure of about 12 feet, surmounted by a lattice mast carrying the main light beacon 40 feet above the sea. These buoys are gradually replacing some of the lightvessels on the British coast, and are named on both sides with the name of the station. They carry a light and fog signal just as the lightvessel they have replaced, but they are unmanned and automatic.

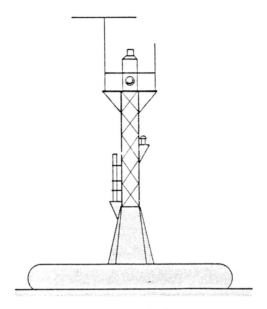

Fig 121   A Lanby buoy

Chapter 10   Use Of Electronic Navigation Instruments

No up-to-date navigation primer would be complete without reference to the wide range of modern electronic instruments now available for fishermen.

While coastal and ocean navigation must depend ultimately upon the basic knowledge acquired, an understanding of electronic navigational aids will enable an appreciation of the safety factors inherent in their use. The UK Department of Trade always consider these as AIDS to navigation and this is emphasised in their examination syllabus whereby basic navigation skills predominate.

Like all equipment dependent upon electrical or battery power, breakdowns can occur, but, as long as you are aware of this and the fact that the reliability of these modern instruments is improving steadily, it is acceptable that they should be used to the greatest extent of their capabilities.

Photographs and brief technical details are given on the following pages of some of the products of the leading manufactures.

Echo Sounders and Fish Finders

There are comprehensive ranges of echo sounders, suitable for all types of small craft and methods of fishing. It is important to obtain professional advice in selecting a suitable model, bearing in mind the performance and depth sounding facilities required.

Some echo sounders use the LED (light emitting diod) display, and others use chart recording modes, while some use both systems including colour displays.

The model illustrated on page 145 is designed to give readings with extreme shallow water accuracy, and is designed for coastal wasters. The compact size makes it the ideal choice for the smaller vessel where space is limited. It provides accurate, non-distorted straight line recordings on 4 inch wide paper, displaying fish and bottom information as to conditions under the hull.

Radio Navigational Systems

A radio navigational receiver gives precise and continuous position fixing in all kinds of weather, thereby rendering it possible always to have safe and efficient navigation. Positional data is given every 10 seconds, with an accuracy down to 50 metres.

Fig 122   Echo sounder and fish finder
By courtesy of Kelvin Hughes Ltd.

Twenty-five transmitter chains cover the waters around Western Europe (except the Mediterranean) and the other chains cover the Indian Ocean, Canada, Japan, the Persian Gulf, NW Australia, and South Africa.

The system constantly receives the analyses beacon signals emitted from land based transmitters. Signals are picked up by the fully automatic receiver via the antenna.

Computations are based on continuous crossbearings which are computed at high speed on two built-in microprocessors, and the results appear in an alpha numeric display. Latitude and longitude are constantly available, expressed in degrees, minutes, and hundredths of a minute.

Autopilots

These may be controlled by gyro or magnetic compass and the main steering compass may be used for autopilot control.

Electronically controlled and reliable they can be fitted with minimal or without structural alterations, and can be changed quickly from auto to manual steering.

A wide range of autopilots are available, the more sophisticated systems have a control unit, course selector, drive unit, distribution unit, rudder

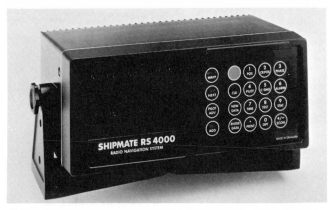

Fig 123   Radio navigator   *By courtesy of Shipmate Navigator Ltd.*

reference unit, together with several options and alternatives. One should always seek expert assistance in selecting the appropriate system for the type of vessel and in the installation.

Illustrated below is an autopilot specifically designed for fishing boats up to 50 feet.

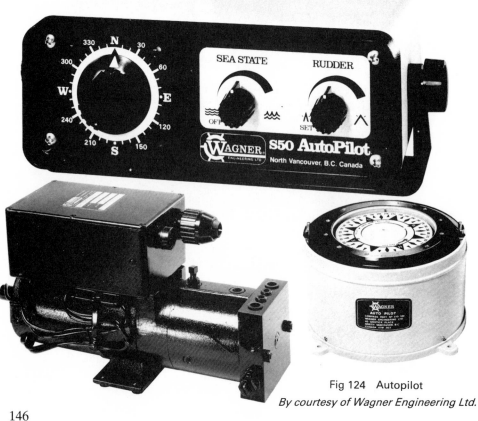

Fig 124   Autopilot

By courtesy of Wagner Engineering Ltd.

A very important factor in the use of autopilots is that the lookout must never be neglected. Unfortunately, in the past, quite a few collisions have taken place due to the neglect of keeping a proper lookout when the vessel is under auto-pilot. A warning system is fitted, which sounds off if a button is not pressed at specified intervals, but, on occasions fishermen have switched off this system and the specified intervals ignored.

Radar

The fitting of radar, and a proper understanding of its uses and capabilities will bring safety and navigational advantages to you. The range of a radar set depends upon the type of set, the height of the scanner aerial, and the atmospheric conditions existing at the time.

Illustrated below is the Racal-Decca 370 set, a high power radar designed specifically for small vessels of every type. It gives a performance on 8 ranges from 0.25 nautical miles up to 48 nautical miles. The aerial is 4 feet wide, and rotates at 25 RPM.

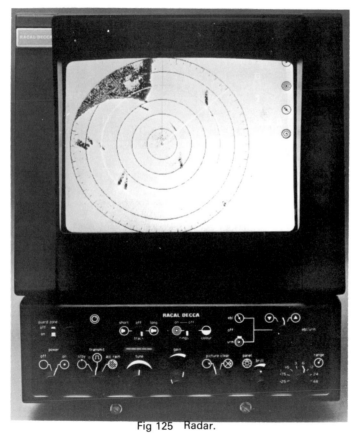

Fig 125   Radar.

By courtesy of Racal-Decca Ltd.

Fig 126   Radar

By courtesy of Kelvin Hughes Ltd.

Another set is the 500 series by Kelvin Hughes is shown here with full details of the controls. It has a similar range, 0.25 to 48 nautical miles, and again is designed specifically for the smaller vessels.

Satellite Navigation

The development of satellite navigation has brought about a major revolution in marine electronic navigational aids. In any part of the World, or in any weather you can now obtain accurate positional fixtures.

Using signals from the US Navy Navigation Satellite System (TRANSIT) satellites which are in transpolar orbits circling the earth every 107 minutes at a height of about 668 miles. Whenever a satellite passes above the horizon of the receiver there exists the opportunity to obtain a postitional fix. Up to 16 good fixes occur in every 24 hour period, and between passes continuous navigational information, such as dead reckoning positions are available.

148

The instrument takes measurements over a period of 10–15 minutes during each satellite pass and can fix a vessel's position within 0.2 of a nautical mile. Every time a satellite passes over, the position is recalculated by the micro-computer. When leaving port the approximate position (within 60 miles) the date, GMT, course and speed is entered.

Various data displays will give the GMT, date, latitude and longitude, also the GMT, date and time of future satellite passes, identification, maximum altitude and direction of rise, the position and time of last fix, and finally given the position of intended destination. Up to 10 intermediate destinations, (way points) together with the tide, speed and direction, the data display will then show the way point number, its range, bearing (true or magnetic), course to steer (great circle or rhumb line), vessel's speed and heading.

Below is a diagram showing the orbital planes of satellites.

Fig 127   Orbital planes of satellites

On page 150 is shown DS4 satellite navigator from Racal-Decca Ltd. and another, the 601S, is from Kelvin Hughes Ltd.

Decca

Possibly the most familiar piece of equipment known to fishermen.

The Decca Navigator is a hyperbolic radio navigation system operated in conjunction with ground transmitter chains working in the 70–130 kHz frequency band. These low frequency signals are transmitted from 51 change of stations, sited in many countries and still expanding.

It is a well established system with the transmitting stations radiating continuous wave signals which the receiver on board, called a decometer, records

149

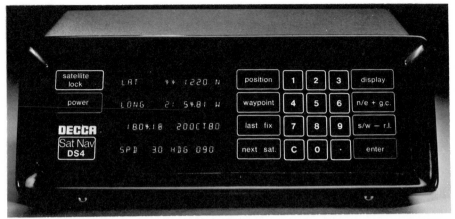

Fig 128 Satellite Navigator. This receiver gives a clear display and providing, simultaneously, latitude longitude, time, date and heading, while other displays of information give all the additional global Sat.-Nav. position fixing data required by the professional fishermen.

By courtesy of Racal-Decca Ltd.

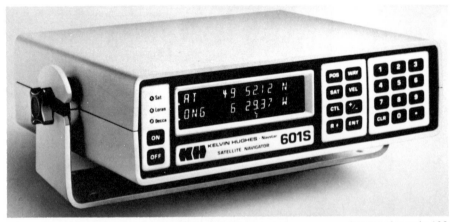

Fig 129 Satellite Navigator. A compact unit it has a static accuracy of approximately 100 yards and the frequency of satellite updated is dependent upon the user's position. At latitude 50° it is every 70 minutes, and at latitude 30° updating will be at intervals of 90 minutes.

By courtesy of Kelvin Hughes Ltd.

on three clock-like dials these readings being referred to a specially latticed navigation chart. The maximum range is around 240 nautical miles within a chain, but greater distances can be obtained according to conditions and with accuracy down to a position within 25 metres.

The special Decca charts required for use with the Decca Navigator can be obtained from any chart supplier, and the receiver is usually rented from the

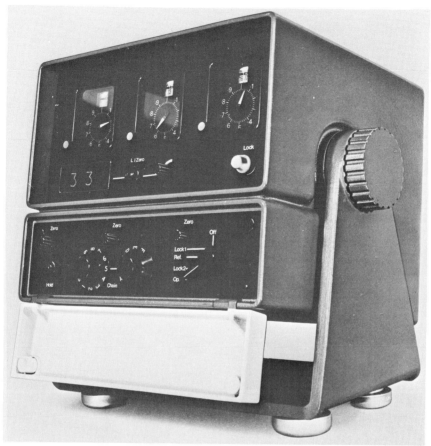

Fig 130   Decca Navigator
By courtesy of Racal-Decca Ltd.

company. The system is easy to learn and simple to operate, while servicing is available to any time and within reach of fishing ports. Above is shown the Decca Navigator Mark 21.

Loran 'C'

The word 'LORAN' is taken from the words LOng RAnge Navigation. Special receiving equipment and lattice charts are required on vessels which use this system.

The Loran 'C' system is a terrestrial based low radio frequency pulse and phase time difference measurement navigation system. It provides medium to long range continuous position fixing, with a precision sufficient to meet coastal and ocean navigation in suitable coverage areas. It is extensively deployed in the United States and Canadian coastal waters, also in other parts of the World including the North Atlantic, the North and Central Pacific, and the Mediterranean Sea.

Omega

Provides continuous position fixing data in the order of two nautical miles suitable for ocean navigation purposes. The system comprises a total of 8 individual stations, located in the USA, Japan, Hawaii, Australia, Reunion, Argentina, Norway, and Liberia, and is operational in the North Atlantic and North Pacific.

Below is an illustration of the Racal-Decca Marine Navigator System MNS 2,000 which is designed to operate from the Decca Navigator, Loran C, Omega, and Transit Satellite System.

All the information given in this chapter must of necessity be brief, fuller details should be obtained from the manufacturers or their representatives or agents at fishing ports before the fitting and operation of electronic navigation equipment.

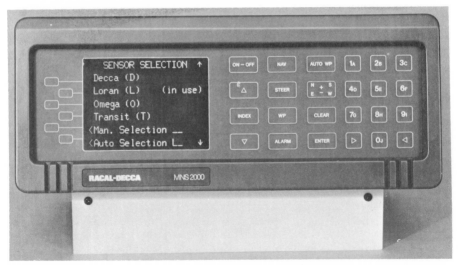

Fig 131   Loran 'C' and Omega Navigator

By courtesy of Racal-Decca Ltd.

Appendix

BIBLIOGRAPHY AND USEFUL ADDRESSES

Bibliography

Fishermen's Handbook by Capt. W. H. Perry (Fishing News Books Ltd.)
Stability and Trim of Fishing Vessels by J. A. Hind (Fishing News Books Ltd.)
The Fishing Cadet's Handbook by A. Hodson (Fishing News Books Ltd.)
International Regulations for Preventing Collisions at Sea (H.M. Stationery Office)
Reed's Almanac by Capt. O. M. Watts, F.R.A.S. (Thomas Reed)
Brown's Nautical Almanac (Brown, Son and Ferguson Ltd.)
Radar Observer's Handbook, Capt. W. Berger (Brown, Son and Ferguson Ltd.)
Home Trade Navigation Guide, Capt. W. MacFarlane (Brown, Son and Ferguson Ltd.)
The Mariner's Highway Code, Capt. D. R. Derrett (Stanford Maritime Limited)
Exercises in Coastal Navigation, Capt. G. W. White (Stanford Maritime Limited)

Some useful addresses

H.M. Coastguard, Sanctuary Buildings, 20, Great Smith Street, London, S.W.1.
H.M. Stationery Office, 49, High Holborn, London, W.C.1.
H.M. Customs and Excise (Headquarters) King's Beam House, Mark Lane, London, E.C.3.
The Mariner's Handbook, Hydrographic Department, Ministry of Defence, Taunton, Somerset.
Meteorological Office, London Weather Centre, 284–6, High Holborn, London, W.C.1.
Registrar General of Shipping and Seamen, Llantrisant Road, Llandaff, Cardiff.
Fishing News Books Ltd., 1 Long Garden Walk, Farnham, Surrey.
Fishing News, Cloister Court, 24 Farringdon Lane, London EC1R 3AV.
Racal Marine Radar Ltd., New Malden, Surrey, KT34 4NR.

TRAVERSE TABLE

332° ↑
208°

↑ 028°
152°

1h 52m

28 DEGREES.

Dist	D.Lat	Dep.	Dist	D.Lat	Dep.	Dist	D.Lat	Dep.	Dist	D.Lat	Dep.	Dist	D.Lat	Dep.
1	00·9	00·5	61	53·9	28·6	121	106·8	56·8	181	159·8	85·0	241	212·8	113·1
2	01·8	00·9	62	54·7	29·1	122	107·7	57·3	182	160·7	85·4	242	213·7	113·6
3	02·6	01·4	63	55·6	29·6	123	108·6	57·7	183	161·6	85·9	243	214·6	114·1
4	03·5	01·9	64	56·5	30·0	124	109·5	58·2	184	162·5	86·4	244	215·4	114·6
5	04·4	02·3	65	57·4	30·5	125	110·4	58·7	185	163·3	86·9	245	216·3	115·0
6	05·3	02·8	66	58·3	31·0	126	111·3	59·2	186	164·2	87·3	246	217·2	115·5
7	06·2	03·3	67	59·2	31·5	127	112·1	59·6	187	165·1	87·8	247	218·1	116·0
8	07·1	03·8	68	60·0	31·9	128	113·0	60·1	188	166·0	88·3	248	219·0	116·4
9	08·0	04·2	69	60·9	32·4	129	113·9	60·6	189	166·9	88·7	249	219·9	116·9
10	08·8	04·7	70	61·8	32·9	130	114·8	61·0	190	167·8	89·2	250	220·7	117·4
11	09·7	05·2	71	62·7	33·3	131	115·7	61·5	191	168·6	89·7	251	221·6	117·9
12	10·6	05·6	72	63·6	33·8	132	116·5	62·0	192	169·5	90·1	252	222·5	118·3
13	11·5	06·1	73	64·5	34·3	133	117·4	62·4	193	170·4	90·6	253	223·4	118·8
14	12·4	06·6	74	65·3	34·7	134	118·3	62·9	194	171·3	91·1	254	224·3	119·2
15	13·2	07·0	75	66·2	35·2	135	119·2	63·4	195	172·2	91·5	255	225·2	119·7
16	14·1	07·5	76	67·1	35·7	136	120·1	63·8	196	173·1	92·0	256	226·0	120·2
17	15·0	08·0	77	68·0	36·1	137	121·0	64·3	197	173·9	92·5	257	226·9	120·7
18	15·9	08·5	78	68·9	36·6	138	121·8	64·8	198	174·8	93·0	258	227·8	121·1
19	16·8	08·9	79	69·8	37·1	139	122·7	65·3	199	175·7	93·4	259	228·7	121·6
20	17·7	09·4	80	70·6	37·6	140	123·6	65·7	200	176·6	93·9	260	229·6	122·1
21	18·5	09·9	81	71·5	38·0	141	124·5	66·2	201	177·5	94·4	261	230·4	122·5
22	19·4	10·3	82	72·4	38·5	142	125·4	66·7	202	178·4	94·8	262	231·3	123·0
23	20·3	10·8	83	73·3	39·0	143	126·3	67·1	203	179·2	95·3	263	232·2	123·5
24	21·2	11·3	84	74·2	39·4	144	127·1	67·6	204	180·1	95·8	264	233·1	123·9
25	22·1	11·7	85	75·1	39·9	145	128·0	68·1	205	181·0	96·2	265	234·0	124·4
26	23·0	12·2	86	75·9	40·4	146	128·9	68·5	206	181·9	96·7	266	234·9	124·9
27	23·8	12·7	87	76·8	40·8	147	129·8	69·0	207	182·8	97·2	267	235·7	125·3
28	24·7	13·1	88	77·7	41·3	148	130·7	69·5	208	183·7	97·7	268	236·6	125·8
29	25·6	13·6	89	78·6	41·8	149	131·6	70·0	209	184·5	98·1	269	237·5	126·3
30	26·5	14·1	90	79·5	42·3	150	132·4	70·4	210	185·4	98·6	270	238·4	126·8
31	27·4	14·6	91	80·3	42·7	151	133·3	70·9	211	186·3	99·1	271	239·3	127·2
32	28·3	15·0	92	81·2	43·2	152	134·2	71·4	212	187·2	99·5	272	240·2	127·7
33	29·1	15·5	93	82·1	43·7	153	135·1	71·8	213	188·1	100·0	273	241·0	128·2
34	30·0	16·0	94	83·0	44·1	154	136·0	72·3	214	189·0	100·5	274	241·9	128·6
35	30·9	16·4	95	83·9	44·6	155	136·9	72·8	215	189·8	100·9	275	242·8	129·1
36	31·8	16·9	96	84·8	45·1	156	137·7	73·2	216	190·7	101·4	276	243·7	129·6
37	32·7	17·4	97	85·6	45·5	157	138·6	73·7	217	191·6	101·9	277	244·6	130·0
38	33·6	17·8	98	86·5	46·0	158	139·5	74·2	218	192·5	102·3	278	245·5	130·5
39	34·4	18·3	99	87·4	46·5	159	140·4	74·6	219	193·4	102·8	279	246·3	131·0
40	35·3	18·8	100	88·3	46·9	160	141·3	75·1	220	194·2	103·3	280	247·2	131·5
41	36·2	19·2	101	89·2	47·4	161	142·2	75·6	221	195·1	103·8	281	248·1	131·9
42	37·1	19·7	102	90·1	47·9	162	143·0	76·1	222	196·0	104·2	282	249·0	132·4
43	38·0	20·2	103	90·9	48·4	163	143·9	76·5	223	196·9	104·7	283	249·9	132·9
44	38·8	20·7	104	91·8	48·8	164	144·8	77·0	224	197·8	105·2	284	250·8	133·3
45	39·7	21·1	105	92·7	49·3	165	145·7	77·5	225	198·7	105·6	285	251·6	133·8
46	40·6	21·6	106	93·6	49·8	166	146·6	77·9	226	199·5	106·1	286	252·5	134·3
47	41·5	22·1	107	94·5	50·2	167	147·5	78·4	227	200·4	106·6	287	253·4	134·7
48	42·4	22·5	108	95·4	50·7	168	148·3	78·9	228	201·3	107·0	288	254·3	135·2
49	43·3	23·0	109	96·2	51·2	169	149·2	79·3	229	202·2	107·5	289	255·2	135·7
50	44·1	23·5	110	97·1	51·6	170	150·1	79·8	230	203·1	108·0	290	256·1	136·1
51	45·0	23·9	111	98·0	52·1	171	151·0	80·3	231	204·0	108·4	291	256·9	136·6
52	45·9	24·4	112	98·9	52·6	172	151·9	80·7	232	204·8	108·9	292	257·8	137·1
53	46·8	24·9	113	99·8	53·1	173	152·7	81·2	233	205·7	109·4	293	258·7	137·6
54	47·7	25·4	114	100·7	53·5	174	153·6	81·7	234	206·6	109·9	294	259·6	138·0
55	48·6	25·8	115	101·5	54·0	175	154·5	82·2	235	207·5	110·3	295	260·5	138·5
56	49·4	26·3	116	102·4	54·5	176	155·4	82·6	236	208·4	110·8	296	261·3	139·0
57	50·3	26·8	117	103·3	54·9	177	156·3	83·1	237	209·3	111·3	297	262·2	139·4
58	51·2	27·2	118	104·2	55·4	178	157·2	83·6	238	210·1	111·7	298	263·1	139·9
59	52·1	27·7	119	105·1	55·9	179	158·0	84·0	239	211·0	112·2	299	264·0	140·4
60	53·0	28·2	120	106·0	56·3	180	158·9	84·5	240	211·9	112·7	300	264·9	140·8

Dist	Dep.	D.Lat	Dist	Dep.	D.Lat	Dist	Dep.	D.Lat	Dist	Dep.	D.Lat	Dist	Dep.	D.Lat

298° ↑
242°

62 DEGREES.

062° ↑
118°

4h 8m

TRAVERSE TABLE
28 DEGREES.

Dist	D.Lat	Dep.	Dist	D.Lat	Dep.	Dist	D.Lat	Dep.	Dist	D.Lat	Dep.	Dist	D.Lat	Dep.
301	265·8	141·3	361	318·7	169·5	421	371·7	197·6	481	424·7	225·8	541	477·7	254·0
302	266·7	141·8	362	319·6	169·9	422	372·6	198·1	482	425·6	226·3	542	478·6	254·5
303	267·5	142·2	363	320·5	170·4	423	373·5	198·6	483	426·5	226·8	543	479·4	254·9
304	268·4	142·7	364	321·4	170·9	424	374·4	199·1	484	427·3	227·2	544	480·3	255·4
305	269·3	143·2	365	322·3	171·4	425	375·3	199·5	485	428·2	227·7	545	481·2	255·9
306	270·2	143·7	366	323·2	171·8	426	376·1	200·0	486	429·1	228·2	546	482·1	256·3
307	271·1	144·1	367	324·0	172·3	427	377·0	200·5	487	430·0	228·6	547	483·0	256·8
308	271·9	144·6	368	324·9	172·8	428	377·9	200·9	488	430·9	229·1	548	483·9	257·3
309	272·8	145·1	369	325·8	173·2	429	378·8	201·4	489	431·8	229·6	549	484·7	257·7
310	273·7	145·5	370	326·7	173·7	430	379·7	201·9	490	432·6	230·0	550	485·6	258·2
311	274·6	146·0	371	327·6	174·2	431	380·6	202·3	491	433·5	230·5	551	486·5	258·7
312	275·5	146·5	372	328·5	174·6	432	381·4	202·8	492	434·4	231·0	552	487·4	259·1
313	276·4	146·9	373	329·3	175·1	433	382·3	203·3	493	435·3	231·4	553	488·3	259·6
314	277·2	147·4	374	330·2	175·6	434	383·2	203·8	494	436·2	231·9	554	489·2	260·1
315	278·1	147·9	375	331·1	176·1	435	384·1	204·2	495	437·1	232·4	555	490·0	260·6
316	279·0	148·4	376	332·0	176·5	436	385·0	204·7	496	437·9	232·9	556	490·9	261·0
317	279·9	148·8	377	332·9	177·0	437	385·8	205·2	497	438·8	233·3	557	491·8	261·5
318	280·8	149·3	378	333·8	177·5	438	386·7	205·6	498	439·7	233·8	558	492·7	262·0
319	281·7	149·8	379	334·6	177·9	439	387·6	206·1	499	440·6	234·3	559	493·6	262·4
320	282·5	150·2	380	335·5	178·4	440	388·5	206·6	500	441·5	234·7	560	494·5	262·9
321	283·4	150·7	381	336·4	178·9	441	389·4	207·0	501	442·4	235·2	561	495·3	263·4
322	284·3	151·2	382	337·3	179·3	442	390·3	207·5	502	443·2	235·7	562	496·2	263·8
323	285·2	151·6	383	338·2	179·8	443	391·1	208·0	503	444·1	236·1	563	497·1	264·3
324	286·1	152·1	384	339·1	180·3	444	392·0	208·4	504	445·0	236·6	564	498·0	264·8
325	287·0	152·6	385	339·9	180·7	445	392·9	208·9	505	445·9	237·1	565	498·9	265·3
326	287·8	153·0	386	340·8	181·2	446	393·8	209·4	506	446·8	237·6	566	499·7	265·7
327	288·7	153·5	387	341·7	181·7	447	394·7	209·9	507	447·7	238·0	567	500·6	266·2
328	289·6	154·0	388	342·6	182·2	448	395·6	210·3	508	448·5	238·5	568	501·5	266·7
329	290·5	154·5	389	343·5	182·6	449	396·4	210·8	509	449·4	239·0	569	502·4	267·1
330	291·4	154·9	390	344·3	183·1	450	397·3	211·3	510	450·3	239·4	570	503·3	267·6
331	292·3	155·4	391	345·2	183·6	451	398·2	211·7	511	451·2	239·9	571	504·2	268·1
332	293·1	155·9	392	346·1	184·0	452	399·1	212·2	512	452·1	240·4	572	505·0	268·5
333	294·0	156·3	393	347·0	184·5	453	400·0	212·7	513	453·0	240·8	573	505·9	269·0
334	294·9	156·8	394	347·9	185·0	454	400·9	213·1	514	453·8	241·3	574	506·8	269·5
335	295·8	157·3	395	348·8	185·4	455	401·7	213·6	515	454·7	241·8	575	507·7	269·9
336	296·7	157·7	396	349·6	185·9	456	402·6	214·1	516	455·6	242·2	576	508·6	270·4
337	297·6	158·2	397	350·5	186·4	457	403·5	214·5	517	456·5	242·7	577	509·5	270·9
338	298·4	158·7	398	351·4	186·8	458	404·4	215·0	518	457·4	243·2	578	510·3	271·4
339	299·3	159·2	399	352·3	187·3	459	405·3	215·5	519	458·2	243·7	579	511·2	271·8
340	300·2	159·6	400	353·2	187·8	460	406·2	216·0	520	459·1	244·1	580	512·1	272·3
341	301·1	160·1	401	354·1	188·3	461	407·0	216·4	521	460·0	244·6	581	513·0	272·8
342	302·0	160·6	402	354·9	188·7	462	407·9	216·9	522	460·9	245·1	582	513·9	273·2
343	302·9	161·0	403	355·8	189·2	463	408·8	217·4	523	461·8	245·5	583	514·8	273·7
344	303·7	161·5	404	356·7	189·7	464	409·7	217·8	524	462·7	246·0	584	515·6	274·2
345	304·6	162·0	405	357·6	190·1	465	410·6	218·3	525	463·5	246·5	585	516·5	274·6
346	305·5	162·4	406	358·5	190·6	466	411·5	218·8	526	464·4	246·9	586	517·4	275·1
347	306·4	162·9	407	359·4	191·1	467	412·3	219·2	527	465·3	247·4	587	518·3	275·4
348	307·3	163·4	408	360·2	191·5	468	413·2	219·7	528	466·2	247·9	588	519·2	276·0
349	308·1	163·8	409	361·1	192·0	469	414·1	220·2	529	467·1	248·4	589	520·1	276·5
350	309·0	164·3	410	362·0	192·5	470	415·0	220·7	530	468·0	248·8	590	520·9	277·0
351	309·9	164·8	411	362·9	193·0	471	415·9	221·1	531	468·8	249·3	591	521·8	277·5
352	310·8	165·3	412	363·8	193·4	472	416·8	221·6	532	469·7	249·8	592	522·7	277·9
353	311·7	165·7	413	364·7	193·9	473	417·6	222·1	533	470·7	250·2	593	523·6	278·4
354	312·6	166·2	414	365·5	194·4	474	418·5	222·5	534	471·5	250·7	594	524·5	278·9
355	313·4	166·7	415	366·4	194·8	475	419·4	223·0	535	472·4	251·2	595	525·4	279·3
356	314·3	167·1	416	367·3	195·3	476	420·3	223·5	536	473·3	251·6	596	526·2	279·8
357	315·2	167·6	417	368·2	195·8	477	421·2	223·9	537	474·1	252·1	597	527·1	280·3
358	316·1	168·1	418	369·1	196·2	478	422·0	224·4	538	475·0	252·6	598	528·0	280·7
359	317·0	168·5	419	370·0	196·7	479	422·9	224·9	539	475·9	253·0	599	528·9	281·2
360	317·9	169·0	420	370·8	197·2	480	423·8	225·3	540	476·8	253·5	600	529·8	281·7

Dist	Dep.	D.Lat	Dist	Dep.	D.Lat	Dist	Dep.	D.Lat	Dist	Dep.	D.Lat	Dist	Dep.	D.Lat

298° ↑
242°
62 DEGREES.
062° ↑
118°   4h 8m
62°

155

M	41°	42°	43°	44°	45°	46°	47°	48°	49°	50°	M
0	2686·24	2766·05	2847·13	2929·55	3013·38	3098·70	3185·59	3274·13	3364·41	3456·53	0
1	2687·56	2767·39	2848·49	2930·93	3014·79	3100·14	3187·05	3275·62	3365·93	3458·08	1
2	2688·88	2768·73	2849·85	2932·32	3016·20	3101·57	3188·51	3277·11	3367·45	3459·64	2
3	2690·20	2770·07	2851·22	2933·71	3017·61	3103·01	3189·97	3278·60	3368·97	3461·19	3
4	2691·52	2771·41	2852·58	2935·09	3019·02	3104·44	3191·44	3280·09	3370·49	3462·74	4
5	2692·84	2772·75	2853·94	2936·48	3020·43	3105·88	3192·90	3281·58	3372·01	3464·29	5
6	2694·16	2774·10	2855·31	2937·87	3021·85	3107·32	3194·36	3283·07	3373·54	3465·85	6
7	2695·49	2775·44	2856·67	2939·26	3023·26	3108·76	3195·83	3284·57	3375·06	3467·40	7
8	2696·81	2776·78	2858·04	2940·64	3024·67	3110·19	3197·29	3286·06	3376·58	3468·96	8
9	2698·13	2778·13	2859·40	2942·03	3026·08	3111·63	3198·76	3287·55	3378·11	3470·52	9
10	2699·45	2779·47	2860·77	2943·42	3027·50	3113·07	3200·23	3289·05	3379·63	3472·07	10
11	2700·78	2780·81	2862·14	2944·81	3028·91	3114·51	3201·69	3290·54	3381·16	3473·63	11
12	2702·10	2782·16	2863·50	2946·20	3030·32	3115·95	3203·16	3292·04	3382·68	3475·19	12
13	2703·42	2783·50	2864·87	2947·59	3031·74	3117·39	3204·63	3293·54	3384·21	3476·75	13
14	2704·75	2784·85	2866·24	2948·98	3033·15	3118·83	3206·10	3295·03	3385·73	3478·30	14
15	2706·07	2786·19	2867·60	2950·37	3034·57	3120·27	3207·56	3296·53	3387·26	3479·86	15
16	2707·40	2787·54	2868·97	2951·76	3035·99	3121·71	3209·03	3298·03	3388·79	3481·42	16
17	2708·72	2788·89	2870·34	2953·15	3037·40	3123·16	3210·50	3299·52	3390·32	3482·98	17
18	2710·05	2790·23	2871·71	2954·55	3038·82	3124·60	3211·97	3301·02	3391·85	3484·54	18
19	2711·38	2791·58	2873·08	2955·94	3040·23	3126·04	3213·44	3302·52	3393·38	3486·11	19
20	2712·70	2792·93	2874·45	2957·33	3041·65	3127·49	3214·91	3304·02	3394·91	3487·67	20
21	2714·03	2794·28	2875·82	2958·73	3043·07	3128·93	3216·38	3305·52	3396·44	3489·23	21
22	2715·36	2795·62	2877·19	2960·12	3044·49	3130·37	3217·86	3307·02	3397·97	3490·79	22
23	2716·68	2796·97	2878·56	2961·51	3045·91	3131·82	3219·33	3308·52	3399·50	3492·36	23
24	2718·01	2798·32	2879·93	2962·91	3047·33	3133·26	3220·80	3310·02	3401·03	3493·92	24
25	2719·34	2799·67	2881·30	2964·30	3048·75	3134·71	3222·27	3311·53	3402·56	3495·49	25
26	2720·67	2801·02	2882·67	2965·70	3050·17	3136·15	3223·75	3313·03	3404·10	3497·05	26
27	2722·00	2802·37	2884·05	2967·09	3051·59	3137·60	3225·22	3314·53	3405·63	3498·62	27
28	2723·33	2803·72	2885·42	2968·49	3053·01	3139·05	3226·69	3316·03	3407·16	3500·18	28
29	2724·66	2805·07	2886·79	2969·89	3054·43	3140·49	3228·17	3317·54	3408·70	3501·75	29
30	2725·99	2806·42	2888·17	2971·28	3055·85	3141·94	3229·64	3319·04	3410·23	3503·32	30
31	2727·32	2807·77	2889·54	2972·68	3057·27	3143·39	3231·12	3320·55	3411·77	3504·89	31
32	2728·65	2809·13	2890·91	2974·08	3058·70	3144·84	3232·60	3322·05	3413·30	3506·45	32
33	2729·98	2810·48	2892·29	2975·48	3060·12	3146·29	3234·07	3323·56	3414·84	3508·02	33
34	2731·31	2811·83	2893·66	2976·88	3061·54	3147·74	3235·55	3325·07	3416·38	3509·59	34
35	2732·64	2813·18	2895·04	2978·28	3062·97	3149·19	3237·03	3326·57	3417·92	3511·16	35
36	2733·97	2814·54	2896·42	2979·68	3064·39	3150·64	3238·51	3328·08	3419·45	3512·73	36
37	2735·31	2815·89	2897·79	2981·08	3065·81	3152·09	3239·98	3329·59	3420·99	3514·30	37
38	2736·64	2817·25	2899·17	2982·48	3067·24	3153·54	3241·46	3331·10	3422·53	3515·88	38
39	2737·97	2818·60	2900·54	2983·88	3068·66	3154·99	3242·94	3332·60	3424·07	3517·45	39
40	2739·30	2819·95	2901·92	2985·28	3070·09	3156·45	3244·42	3334·11	3425·61	3519·02	40
41	2740·64	2821·31	2903·30	2986·68	3071·52	3157·90	3245·90	3335·62	3427·15	3520·60	41
42	2741·97	2822·67	2904·68	2988·08	3072·94	3159·35	3247·38	3337·13	3428·70	3522·17	42
43	2743·31	2824·02	2906·06	2989·48	3074·37	3160·81	3248·87	3338·65	3430·24	3523·75	43
44	2744·64	2825·38	2907·43	2990·88	3075·80	3162·26	3250·35	3340·16	3431·78	3525·32	44
45	2745·98	2826·73	2908·81	2992·29	3077·23	3163·71	3251·83	3341·67	3433·32	3526·90	45
46	2747·31	2828·09	2910·19	2993·69	3078·66	3165·17	3253·31	3343·18	3434·87	3528·47	46
47	2748·65	2829·45	2911·57	2995·09	3080·09	3166·62	3254·80	3344·69	3436·41	3530·05	47
48	2749·98	2830·81	2912·95	2996·50	3081·52	2168·08	3256·28	3346·21	3437·95	3531·63	48
49	2751·32	2832·16	2914·33	2997·90	3082·95	3169·54	3257·77	3347·72	3439·50	3533·21	49
50	2752·66	2833·52	2915·72	2999·31	3084·38	3170·99	3259·25	3349·24	3441·05	3534·79	50
51	2754·00	2834·88	2917·10	3000·71	3085·81	3172·45	3260·74	3350·75	3442·59	3536·37	51
52	2755·33	2836·24	2918·48	3002·12	3087·24	3173·91	3262·22	3352·27	3444·14	3537·95	52
53	2756·67	2837·60	2919·86	3003·53	3088·67	3175·37	3263·71	3353·78	3445·69	3539·53	53
54	2758·01	2838·96	2921·24	3004·93	3090·10	3176·83	3265·20	3355·30	3447·23	3541·11	54
55	2759·35	2840·32	2922·63	3006·34	3091·53	3178·28	3266·68	3356·82	3448·78	3542·69	55
56	2760·69	2841·68	2924·01	3007·75	3092·97	3179·74	3268·17	3358·33	3450·33	3544·27	56
57	2762·03	2843·04	2925·39	3009·16	3094·40	3181·20	3269·66	3359·85	3451·88	3545·85	57
58	2763·37	2844·40	2926·78	3010·56	3095·83	3182·66	3271·15	3361·37	3453·43	3547·44	58
59	2764·71	2845·77	2928·16	3011·97	3097·27	3184·13	3272·64	3362·89	3454·98	3549·02	59
60	2766·05	2847·13	2929·55	3013·38	3098·70	3185·59	3274·13	3364·41	3456·53	3550·60	60
M	41°	42°	43°	44°	45°	46°	47°	48°	49°	50°	M

(Terrestrial Spheroid) MERIDIONAL PARTS. Compression $\frac{1}{297.465}$

M	51°	52°	53°	54°	55°	56°	57°	58°	59°	60°	M
0	3550·60	3646·74	3745·05	3845·69	3948·78	4054·48	4162·97	4274·43	4389·06	4507·08	0
1	3552·19	3648·36	3746·71	3847·38	3950·52	4056·27	4164·81	4276·31	4391·00	4509·07	1
2	3553·77	3649·98	3748·37	3849·08	3952·26	4058·05	4166·64	4278·20	4392·94	4511·07	2
3	3555·36	3651·60	3750·03	3850·78	3954·00	4059·84	4168·47	4280·08	4394·88	4513·07	3
4	3556·95	3653·22	3751·69	3852·48	3955·74	4061·63	4170·31	4281·97	4396·82	4515·07	4
5	3558·53	3654·84	3753·35	3854·18	3957·48	4063·41	4172·14	4283·86	4398·76	4517·07	5
6	3560·12	3656·47	3755·01	3855·88	3959·23	4065·20	4173·98	4285·75	4400·70	4519·08	6
7	3561·71	3658·09	3756·67	3857·58	3960·97	4066·99	4175·82	4287·64	4402·65	4521·08	7
8	3563·30	3659·72	3758·33	3859·29	3962·72	4068·78	4177·66	4289·53	4404·59	4523·08	8
9	3564·89	3661·34	3760·00	3860·99	3964·46	4070·57	4179·50	4291·42	4406·54	4525·09	9
10	3566·48	3662·97	3761·66	3862·69	3966·21	4072·37	4181·34	4293·31	4408·49	4527·09	10
11	3568·07	3664·59	3763·33	3864·40	3967·96	4074·16	4183·18	4295·20	4410·43	4529·10	11
12	3569·66	3666·22	3764·99	3866·10	3969·70	4075·95	4185·02	4297·09	4412·38	4531·11	12
13	3571·25	3667·85	3766·66	3867·81	3971·45	4077·75	4186·86	4298·99	4414·33	4533·12	13
14	3572·85	3669·48	3768·32	3869·52	3973·20	4079·54	4188·71	4300·89	4416·28	4535·13	14
15	3574·44	3671·11	3769·99	3871·22	3974·95	4081·34	4190·55	4302·78	4418·24	4537·14	15
16	3576·03	3672·74	3771·66	3872·93	3976·70	4083·13	4192·40	4304·68	4420·19	4539·15	16
17	3577·63	3674·37	3773·33	3874·64	3978·46	4084·93	4194·24	4306·58	4422·14	4541·17	17
18	3579·22	3676·00	3774·99	3876·35	3980·21	4086·73	4196·09	4308·48	4424·10	4543·18	18
19	3580·82	3677·63	3776·66	3878·06	3981·96	4088·53	4197·94	4310·38	4426·05	4545·20	19
20	3582·41	3679·26	3778·33	3879·77	3983·71	4090·33	4199·79	4312·28	4428·01	4547·21	20
21	3584·01	3680·89	3780·00	3881·48	3985·47	4092·13	4201·64	4314·18	4429·97	4549·23	21
22	3585·61	3682·53	3781·68	3883·19	3987·22	4093·93	4203·49	4316·08	4431·93	4551·25	22
23	3587·20	3684·16	3783·35	3884·91	3988·98	4095·73	4205·34	4317·98	4433·89	4553·27	23
24	3588·80	3685·79	3785·02	3886·62	3990·74	4097·54	4207·19	4319·89	4435·85	4555·29	24
25	3590·40	3687·43	3786·69	3888·33	3992·49	4099·34	4209·04	4321·79	4437·81	4557·31	25
26	3592·00	3689·07	3788·37	3890·05	3994·25	4101·14	4210·90	4323·70	4439·77	4559·33	26
27	3593·60	3690·70	3790·04	3891·76	3996·01	4102·95	4212·75	4325·61	4441·73	4561·36	27
28	3595·20	3692·34	3791·72	3893·48	3997·77	4104·75	4214·61	4327·52	4443·70	4563·38	28
29	3596·80	3693·98	3793·40	3895·20	3999·53	4106·56	4216·46	4329·42	4445·66	4565·41	29
30	3598·40	3695·61	3795·07	3896·91	4001·29	4108·37	4218·32	4331·33	4447·63	4567·44	30
31	3600·01	3697·25	3796·75	3898·63	4003·06	4110·18	4220·18	4333·24	4449·60	4569·46	31
32	3601·61	3698·89	3798·43	3900·35	4004·82	4111·99	4222·04	4335·16	4451·56	4571·49	32
33	3603·21	3700·53	3800·11	3902·07	4006·58	4113·80	4223·90	4337·07	4453·53	4573·52	33
34	3604·82	3702·17	3801·79	3903·79	4008·35	4115·61	4225·76	4338·98	4455·50	4575·55	34
35	3606·42	3703·82	3803·47	3905·51	4010·11	4117·42	4227·62	4340·90	4457·48	4577·59	35
36	3608·03	3705·46	3805·15	3907·24	4011·88	4119·23	4229·48	4342·81	4459·45	4579·62	36
37	3609·63	3707·10	3806·83	3908·96	4013·64	4121·05	4231·34	4344·73	4461·42	4581·65	37
38	3611·24	3708·74	3808·51	3910·68	4015·41	4122·86	4233·21	4346·65	4463·40	4583·69	38
39	3612·85	3710·39	3810·19	3912·41	4017·18	4124·67	4235·07	4348·56	4465·37	4585·72	39
40	3614·46	3712·03	3811·88	3914·13	4018·95	4126·49	4236·94	4350·48	4467·35	4587·76	40
41	3616·06	3713·68	3813·56	3915·86	4020·72	4128·31	4238·80	4352·40	4469·32	4589·80	41
42	3617·67	3715·32	3815·25	3917·58	4022·49	4130·12	4240·67	4354·32	4471·30	4591·84	42
43	3619·28	3716·97	3816·93	3919·31	4024·26	4131·94	4242·54	4356·25	4473·28	4593·88	43
44	3620·89	3718·62	3818·62	3921·04	4026·03	4133·76	4244·41	4358·17	4475·26	4595·92	44
45	3622·50	3720·26	3820·30	3922·77	4027·80	4135·58	4246·28	4360·09	4477·24	4597·96	45
46	3624·12	3721·91	3821·99	3924·50	4029·58	4137·40	4248·15	4362·02	4479·22	4600·01	46
47	3625·73	3723·56	3823·68	3926·23	4031·35	4139·22	4250·02	4363·94	4481·21	4602·05	47
48	3627·34	3725·21	3825·37	3927·96	4033·13	4141·04	4251·89	4365·87	4483·19	4604·10	48
49	3628·95	3726·86	3827·06	3929·69	4034·90	4142·87	4253·77	4367·80	4485·18	4606·15	49
50	3630·57	3728·51	3828·75	3931·42	4036·68	4144·69	4255·64	4369·72	4487·16	4608·19	50
51	3632·18	3730·16	3830·44	3933·15	4038·45	4146·52	4257·52	4371·65	4489·15	4610·24	51
52	3633·80	3731·81	3832·13	3934·88	4040·23	4148·34	4259·39	4373·58	4491·14	4612·29	52
53	3635·41	3733·47	3833·82	3936·62	4042·01	4150·17	4261·27	4375·51	4493·13	4614·34	53
54	3637·03	3735·12	3835·52	3938·35	4043·79	4151·99	4263·15	4377·45	4495·12	4616·40	54
55	3638·64	3736·77	3837·21	3940·09	4045·57	4153·82	4265·02	4379·38	4497·11	4618·45	55
56	3640·26	3738·43	3838·90	3941·83	4047·35	4155·65	4266·90	4381·31	4499·10	4620·50	56
57	3641·88	3740·08	3840·60	3943·56	4049·13	4157·48	4268·78	4383·25	4501·09	4622·56	57
58	3643·50	3741·74	3842·29	3945·30	4050·92	4159·31	4270·67	4385·18	4503·09	4624·62	58
59	3645·12	3743·40	3843·99	3947·04	4052·70	4161·14	4272·55	4387·12	4505·08	4626·67	59
60	3646·74	3745·05	3845·69	3948·78	4054·48	4162·97	4274·43	4389·06	4507·08	4628·73	60
M	51°	52°	53°	54°	55°	56°	57°	58°	59°	60°	M

LOGARITHMS

No. 1000——1599   Log. 00000——20385

Angles 2·	Angles 3·	No.	0	1	2	3	4	5	6	7	8	9	D
° ′	° ′												
1 40	16 40	100	00000	00043	00087	00130	00173	00217	00260	00303	00346	00389	43
1 41	16 50	101	00432	00475	00518	00561	00604	00647	00689	00732	00775	00817	43
1 42	17 00	102	00860	00903	00945	00988	01030	01072	01115	01157	01199	01242	42
1 43	17 10	103	01284	01326	01368	01410	01452	01494	01536	01578	01620	01662	42
1 44	17 20	104	01703	01745	01787	01828	01870	01912	01953	01995	02036	02078	42
1 45	17 30	105	02119	02160	02202	02243	02284	02325	02366	02408	02449	02490	41
1 46	17 40	106	02531	02572	02612	02653	02694	02735	02776	02816	02857	02898	41
1 47	17 50	107	02938	02979	03020	03060	03100	03141	03181	03222	03262	03302	40
1 48	18 00	108	03342	03383	03423	03463	03503	03543	03583	03623	03663	03703	40
1 49	18 10	109	03743	03783	03822	03862	03902	03941	03981	04021	04060	04100	40
1 50	18 20	110	04139	04179	04218	04258	04297	04336	04376	04415	04454	04493	39
1 51	18 30	111	04532	04571	04611	04650	04689	04728	04766	04805	04844	04883	39
1 52	18 40	112	04922	04961	04999	05038	05077	05115	05154	05192	05231	05269	39
1 53	18 50	113	05308	05346	05385	05423	05461	05500	05538	05576	05614	05652	38
1 54	19 00	114	05691	05729	05767	05805	05843	05881	05919	05956	05994	06032	38
1 55	19 10	115	06070	06108	06145	06183	06221	06258	06296	06333	06371	06408	38
1 56	19 20	116	06446	06483	06521	06558	06595	06633	06670	06707	06744	06781	37
1 57	19 30	117	06819	06856	06893	06930	06967	07004	07041	07078	07115	07151	37
1 58	19 40	118	07188	07225	07262	07299	07335	07372	07409	07445	07482	07518	37
1 59	19 50	119	07555	07591	07628	07664	07700	07737	07773	07809	07846	07882	36
2 00	20 00	120	07918	07954	07990	08027	08063	08099	08135	08171	08207	08243	36
2 01	20 10	121	08279	08314	08350	08386	08422	08458	08493	08529	08565	08600	36
2 02	20 20	122	08636	08672	08707	08743	08778	08814	08849	08885	08920	08955	36
2 03	20 30	123	08991	09026	09061	09096	09132	09167	09202	09237	09272	09307	35
2 04	20 40	124	09342	09377	09412	09447	09482	09517	09552	09587	09622	09656	35
2 05	20 50	125	09691	09726	09760	09795	09830	09864	09899	09934	09968	10003	35
2 06	21 00	126	10037	10072	10106	10140	10175	10209	10243	10278	10312	10346	34
2 07	21 10	127	10380	10415	10449	10483	10517	10551	10585	10619	10653	10687	34
2 08	21 20	128	10721	10755	10789	10823	10857	10890	10924	10958	10992	11025	34
2 09	21 30	129	11059	11093	11126	11160	11193	11227	11261	11294	11328	11361	34
2 10	21 40	130	11394	11428	11461	11494	11528	11561	11594	11628	11661	11694	33
2 11	21 50	131	11727	11760	11793	11827	11860	11893	11926	11959	11992	12025	33
2 12	22 00	132	12057	12090	12123	12156	12189	12222	12254	12287	12320	12353	33
2 13	22 10	133	12385	12418	12450	12483	12516	12548	12581	12613	12646	12678	33
2 14	22 20	134	12711	12743	12775	12808	12840	12872	12905	12937	12969	13001	32
2 15	22 30	135	13033	13066	13098	13130	13162	13194	13226	13258	13290	13322	32
2 16	22 40	136	13354	13386	13418	13450	13481	13513	13545	13577	13609	13640	32
2 17	22 50	137	13672	13704	13735	13767	13799	13830	13862	13893	13925	13956	32
2 18	23 00	138	13988	14019	14051	14082	14114	14145	14176	14208	14239	14270	31
2 19	23 10	139	14302	14333	14364	14395	14426	14457	14489	14520	14551	14582	31
2 20	23 20	140	14613	14644	14675	14706	14737	14768	14799	14829	14860	14891	31
2 21	23 30	141	14922	14953	14984	15014	15045	15076	15106	15137	15168	15198	31
2 22	23 40	142	15229	15259	15290	15321	15351	15382	15412	15442	15473	15503	31
2 23	23 50	143	15534	15564	15594	15625	15655	15685	15715	15746	15776	15806	30
2 24	24 00	144	15836	15866	15897	15927	15957	15987	16017	16047	16077	16107	30
2 25	24 10	145	16137	16167	16197	16227	16256	16286	16316	16346	16376	16406	30
2 26	24 20	146	16435	16465	16495	16524	16554	16584	16613	16643	16673	16702	30
2 27	24 30	147	16732	16761	16791	16820	16850	16879	16909	16938	16967	16997	30
2 28	24 40	148	17026	17056	17085	17114	17143	17173	17202	17231	17260	17290	29
2 29	24 50	149	17319	17348	17377	17406	17435	17464	17493	17522	17551	17580	29
2 30	25 00	150	17609	17638	17667	17696	17725	17754	17783	17811	17840	17869	29
2 31	25 10	151	17898	17926	17955	17984	18013	18041	18070	18099	18127	18156	29
2 32	25 20	152	18184	18213	18242	18270	18299	18327	18355	18384	18412	18441	29
2 33	25 30	153	18469	18498	18526	18554	18583	18611	18639	18667	18696	18724	28
2 34	25 40	154	18752	18780	18808	18837	18865	18893	18921	18949	18977	19005	28
2 35	25 50	155	19033	19061	19089	19117	19145	19173	19201	19229	19257	19285	28
2 36	26 00	156	19313	19340	19368	19396	19424	19451	19479	19507	19535	19562	28
2 37	26 10	157	19590	19618	19645	19673	19701	19728	19756	19783	19811	19838	28
2 38	26 20	158	19866	19893	19921	19948	19976	20003	20030	20058	20085	20112	27
2 39	26 30	159	20140	20167	20194	20222	20249	20276	20303	20331	20358	20385	27

| Angles | | No. | 0 | 1 | 2 | 3 | 4 | 5 | 6 | 7 | 8 | 9 | D. |

For fifth-figure differences see page 174.

158

LOGARITHMS

Angles 2·	Angles 3·	No.	0	1	2	3	4	5	6	7	8	9	D.
° ′	° ′												
2 40	26 40	160	20412	20439	20466	20493	20520	20548	20575	20602	20629	20656	27
2 41	26 50	161	20683	20710	20737	20763	20790	20817	20844	20871	20898	20925	27
2 42	27 00	162	20952	20978	21005	21032	21059	21085	21112	21139	21165	21192	27
2 43	27 10	163	21219	21245	21272	21299	21325	21352	21378	21405	21431	21458	27
2 44	27 20	164	21484	21511	21537	21564	21590	21617	21643	21669	21696	21722	26
2 45	27 30	165	21748	21775	21801	21827	21854	21880	21906	21932	21958	21985	26
2 46	27 40	166	22011	22037	22063	22089	22115	22141	22168	22194	22220	22246	26
2 47	27 50	167	22272	22298	22324	22350	22376	22402	22427	22453	22479	22505	26
2 48	28 00	168	22531	22557	22583	22608	22634	22660	22686	22712	22737	22763	26
2 49	28 10	169	22789	22814	22840	22866	22891	22917	22943	22968	22994	23019	26
2 50	28 20	170	23045	23070	23096	23122	23147	23172	23198	23223	23249	23274	26
2 51	28 30	171	23300	23325	23350	23376	23401	23426	23452	23477	23502	23528	25
2 52	28 40	172	23553	23578	23603	23629	23654	23679	23704	23729	23754	23780	25
2 53	28 50	173	23805	23830	23855	23880	23905	23930	23955	23980	24005	24030	25
2 54	29 00	174	24055	24080	24105	24130	24155	24180	24204	24229	24254	24279	25
2 55	29 10	175	24304	24329	24353	24378	24403	24428	24452	24477	24502	24527	25
2 56	29 20	176	24551	24576	24601	24625	24650	24675	24699	24724	24748	24773	25
2 57	29 30	177	24797	24822	24846	24871	24895	24920	24944	24969	24993	25018	25
2 58	29 40	178	25042	25066	25091	25115	25140	25164	25188	25213	25237	25261	24
2 59	29 50	179	25285	25310	25334	25358	25382	25406	25431	25455	25479	25503	24
3 00	30 00	180	25527	25551	25576	25600	25624	25648	25672	25696	25720	25744	24
3 01	30 10	181	25768	25792	25816	25840	25864	25888	25912	25936	25959	25983	24
3 02	30 20	182	26007	26031	26055	26079	26103	26126	26150	26174	26198	26221	24
3 03	30 30	183	26245	26269	26293	26316	26340	26364	26387	26411	26435	26458	24
3 04	30 40	184	26482	26505	26529	26553	26576	26600	26623	26647	26670	26694	24
3 05	30 50	185	26717	26741	26764	26788	26811	26834	26858	26881	26905	26928	23
3 06	31 00	186	26951	26975	26998	27021	27045	27068	27091	27114	27138	27161	23
3 07	31 10	187	27184	27207	27231	27254	27277	27300	27323	27346	27370	27393	23
3 08	31 20	188	27416	27439	27462	27485	27508	27531	27554	27577	27600	27623	23
3 09	31 30	189	27646	27669	27692	27715	27738	27761	27784	27807	27830	27853	23
3 10	31 40	190	27875	27898	27921	27944	27967	27990	28012	28035	28058	28081	23
3 11	31 50	191	28103	28126	28149	28172	28194	28217	28240	28262	28285	28308	23
3 12	32 00	192	28330	28353	28375	28398	28421	28443	28466	28488	28511	28533	23
3 13	32 10	193	28556	28578	28601	28623	28646	28668	28691	28713	28735	28758	23
3 14	32 20	194	28780	28803	28825	28847	28870	28892	28914	28937	28959	28981	22
3 15	32 30	195	29004	29026	29048	29070	29093	29115	29137	29159	29181	29203	22
3 16	32 40	196	29226	29248	29270	29292	29314	29336	29358	29380	29403	29425	22
3 17	32 50	197	29447	29469	29491	29513	29535	29557	29579	29601	29623	29645	22
3 18	33 00	198	29667	29688	29710	29732	29754	29776	29798	29820	29842	29864	22
3 19	33 10	199	29885	29907	29929	29951	29973	29994	30016	30038	30060	30081	22
3 20	33 20	200	30103	30125	30146	30168	30190	30211	30233	30255	30276	30298	22
3 21	33 30	201	30320	30341	30363	30384	30406	30428	30449	30471	30492	30514	22
3 22	33 40	202	30535	30557	30578	30600	30621	30643	30664	30685	30707	30728	22
3 23	33 50	203	30750	30771	30792	30814	30835	30856	30878	30899	30920	30942	21
3 24	34 00	204	30963	30984	31006	31027	31048	31069	31091	31112	31133	31154	21
3 25	34 10	205	31175	31197	31218	31239	31260	31281	31302	31323	31345	31366	21
3 26	34 20	206	31387	31408	31429	31450	31471	31492	31513	31534	31555	31576	21
3 27	34 30	207	31597	31618	31639	31660	31681	31702	31723	31744	31765	31785	21
3 28	34 40	208	31806	31827	31848	31869	31890	31911	31931	31952	31973	31994	21
3 29	34 50	209	32015	32035	32056	32077	32098	32118	32139	32160	32181	32201	21
3 30	35 00	210	32222	32243	32263	32284	32305	32325	32346	32367	32387	32408	21
3 31	35 10	211	32428	32449	32469	32490	32511	32531	32552	32572	32593	32613	21
3 32	35 20	212	32634	32654	32675	32695	32716	32736	32756	32777	32797	32818	20
3 33	35 30	213	32838	32858	32879	32899	32919	32940	32960	32981	33001	33021	20
3 34	35 40	214	33041	33062	33082	33102	33123	33143	33163	33183	33203	33224	20
3 35	35 50	215	33244	33264	33284	33304	33325	33345	33365	33385	33405	33425	20
3 36	36 00	216	33445	33466	33486	33506	33526	33546	33566	33586	33606	33626	20
3 37	36 10	217	33646	33666	33686	33706	33726	33746	33766	33786	33806	33826	20
3 38	36 20	218	33846	33866	33886	33905	33925	33945	33965	33985	34005	34025	20
3 39	36 30	219	34044	34064	34084	34104	34124	34144	34163	34183	34203	34223	20
Angles		No.	0	1	2	3	4	5	6	7	8	9	D.

For fifth-figure differences see page 174.

LOGS. OF TRIG. FUNCTIONS

	Sine	Parts	Cosec.	Tan.	Parts	Cotan.	Secant	Parts	Cosine	
00·0	9·23967	′	10·76033	9·24632	′	10·75368	10·00665	′	9·99335	60′
01·0	9·24039	·1 7	10·75961	9·24706	·1 7	10·75294	10·00667	·1 0	9·99333	
02·0	9·24110	·2 14	10·75890	9·24779	·2 15	10·75221	10·00669	·2 0	9·99331	
03·0	9·24181	·3 21	10·75819	9·24853	·3 22	10·75147	10·00672	·3 1	9·99328	
04·0	9·24253	·4 28	10·75747	9·24926	·4 29	10·75074	10·00674	·4 1	9·99326	
05·0	9·24324	·5 36	10·75676	9·25000	·5 37	10·75000	10·00676	·5 1	9·99324	55′
06·0	9·24395	·6 43	10·75605	9·25073	·6 44	10·74927	10·00678	·6 1	9·99322	
07·0	9·24466	·7 50	10·75534	9·25146	·7 51	10·74854	10·00681	·7 2	9·99320	
08·0	9·24536	·8 57	10·75464	9·25219	·8 59	10·74781	10·00683	·8 2	9·99317	
09·0	9·24607	·9 64	10·75393	9·25292	·9 66	10·74708	10·00685	·9 2	9·99315	
10·0	9·24678		10·75323	9·25365		10·74635	10·00687		9·99313	50′
11·0	9·24748	·1 7	10·75252	9·25437	·1 7	10·74563	10·00690	·1 0	9·99310	
12·0	9·24818	·2 14	10·75182	9·25510	·2 14	10·74490	10·00692	·2 0	9·99308	
13·0	9·24888	·3 21	10·75112	9·25582	·3 22	10·74418	10·00694	·3 1	9·99306	
14·0	9·24958	·4 28	10·75042	9·25655	·4 29	10·74345	10·00696	·4 1	9·99304	
15·0	9·25028	·5 35	10·74972	9·25727	·5 36	10·74273	10·00699	·5 1	9·99301	45′
16·0	9·25098	·6 42	10·74902	9·25799	·6 43	10·74201	10·00701	·6 1	9·99299	
17·0	9·25168	·7 49	10·74832	9·25871	·7 50	10·74129	10·00703	·7 2	9·99297	
18·0	9·25237	·8 56	10·74763	9·25943	·8 58	10·74057	10·00706	·8 2	9·99294	
19·0	9·25307	·9 63	10·74693	9·26015	·9 65	10·73985	10·00708	·9 2	9·99292	
20·0	9·25376		10·74624	9·26086		10·73914	10·00710		9·99290	40′
21·0	9·25445	·1 7	10·74555	9·26158	·1 7	10·73842	10·00713	·1 0	9·99288	
22·0	9·25514	·2 14	10·74486	9·26229	·2 14	10·73771	10·00715	·2 0	9·99285	
23·0	9·25583	·3 21	10·74417	9·26301	·3 21	10·73700	10·00717	·3 1	9·99283	
24·0	9·25652	·4 27	10·74348	9·26372	·4 28	10·73628	10·00719	·4 1	9·99281	
25·0	9·25721	·5 34	10·74279	9·26443	·5 36	10·73557	10·00722	·5 1	9·99278	35′
26·0	9·25790	·6 41	10·74210	9·26514	·6 43	10·73486	10·00724	·6 1	9·99276	
27·0	9·25858	·7 48	10·74142	9·26585	·7 50	10·73415	10·00727	·7 2	9·99274	
28·0	9·25927	·8 55	10·74073	9·26656	·8 57	10·73345	10·00729	·8 2	9·99271	
29·0	9·25995	·9 62	10·74005	9·26726	·9 64	10·73274	10·00731	·9 2	9·99269	
30·0	9·26063		10·73937	9·26797		10·73203	10·00733		9·99267	30′
31·0	9·26131	·1 7	10·73869	9·26867	·1 7	10·73133	10·00736	·1 0	9·99264	
32·0	9·26199	·2 14	10·73801	9·26938	·2 14	10·73063	10·00738	·2 0	9·99262	
33·0	9·26267	·3 20	10·73733	9·27008	·3 21	10·72992	10·00740	·3 1	9·99260	
34·0	9·26335	·4 27	10·73665	9·27078	·4 28	10·72922	10·00743	·4 1	9·99257	
35·0	9·26403	·5 34	10·73597	9·27148	·5 35	10·72852	10·00745	·5 1	9·99255	25′
36·0	9·26470	·6 41	10·73530	9·27218	·6 42	10·72782	10·00748	·6 1	9·99253	
37·0	9·26538	·7 47	10·73462	9·27288	·7 49	10·72712	10·00750	·7 2	9·99250	
38·0	9·26605	·8 54	10·73395	9·27357	·8 56	10·72643	10·00752	·8 2	9·99248	
39·0	9·26672	·9 61	10·73328	9·27427	·9 63	10·72573	10·00755	·9 2	9·99245	
40·0	9·26740		10·73261	9·27496		10·72504	10·00757		9·99243	20′
41·0	9·26807	·1 7	10·73193	9·27566	·1 7	10·72434	10·00759	·1 0	9·99241	
42·0	9·26873	·2 13	10·73127	9·27635	·2 14	10·72365	10·00762	·2 0	9·99238	
43·0	9·26940	·3 20	10·73060	9·27704	·3 21	10·72296	10·00764	·3 1	9·99236	
44·0	9·27007	·4 27	10·72993	9·27773	·4 28	10·72227	10·00767	·4 1	9·99234	
45·0	9·27074	·5 33	10·72927	9·27842	·5 35	10·72158	10·00769	·5 1	9·99231	15′
46·0	9·27140	·6 40	10·72860	9·27911	·6 41	10·72089	10·00771	·6 1	9·99229	
47·0	9·27206	·7 47	10·72794	9·27980	·7 48	10·72020	10·00774	·7 2	9·99226	
48·0	9·27273	·8 53	10·72727	9·28049	·8 55	10·71951	10·00776	·8 2	9·99224	
49·0	9·27339	·9 60	10·72661	9·28117	·9 62	10·71883	10·00779	·9 2	9·99221	
50·0	9·27405		10·72595	9·28186		10·71814	10·00781		9·99219	10′
51·0	9·27471	·1 7	10·72529	9·28254	·1 7	10·71746	10·00783	·1 0	9·99217	
52·0	9·27537	·2 13	10·72463	9·28323	·2 14	10·71678	10·00786	·2 0	9·99214	
53·0	9·27603	·3 20	10·72398	9·28391	·3 20	10·71609	10·00788	·3 1	9·99212	
54·0	9·27668	·4 26	10·72332	9·28459	·4 27	10·71541	10·00791	·4 1	9·99209	
55·0	9·27734	·5 33	10·72266	9·28527	·5 34	10·71473	10·00793	·5 1	9·99207	5′
56·0	9·27799	·6 39	10·72201	9·28595	·6 41	10·71405	10·00796	·6 1	9·99204	
57·0	9·27865	·7 46	10·72136	9·28662	·7 48	10·71338	10·00798	·7 2	9·99202	
58·0	9·27930	·8 52	10·72070	9·28730	·8 54	10·71270	10·00800	·8 2	9·99200	
59·0	9·27995	·9 59	10·72005	9·28798	·9 61	10·71202	10·00803	·9 2	9·99197	
60·0	9·28060		10·71940	9·28865		10·71135	10·00805		9·99195	0′

169°
349°

′	Sine	Parts		Cosec.	Tan.	Parts		Cotan.	Secant	Parts		Cosine	
00·0	9·85693	′		10·14307	10·01516	′		9·98484	10·15823	′		9·84177	60′
01·0	9·85706			10·14294	10·01542			9·98458	10·15836			9·84164	
02·0	9·85718	·1	1	10·14282	10·01567	·1	3	9·98433	10·15849	·1	1	9·84151	
03·0	9·85730			10·14270	10·01592			9·98408	10·15862			9·84138	
04·0	9·85742			10·14258	10·01617			9·98383	10 15875			9 84125	
05·0	9·85754	2	2	10·14246	10·01643	2	5	9·98357	10·15888	·2	3	9·84112	55′
06·0	9·85767			10·14234	10·01668			9·98332	10·15902			9·84099	
07·0	9·85779			10·14221	10·01693			9·98307	10·15915			9·84085	
08·0	9·85791	·3	4	10·14209	10·01719	·3	8	9·98281	10·15928	·3	4	9·84072	
09·0	9·85803			10·14197	10·01744			9·98256	10·15941			9·84059	
10·0	9·85815			10·14185	10·01769			9·98231	10·15954			9·84046	50′
11·0	9·85827	·4	5	10·14173	10·01794	·4	10	9·98206	10·15967	·4	5	9·84033	
12·0	9·85839			10·14161	10·01820			9·98180	10·15980			9·84020	
13·0	9·85851			10·14149	10·01845			9·98155	10·15994			9·84006	
14·0	9·85864	·5	6	10·14137	10·01870	·5	13	9·98130	10·16007	·5	7	9·83993	
15·0	9·85876			10·14124	10·01896			9·98104	10·16020			9·83980	45′
16·0	9·85888			10·14112	10·01921			9·98079	10·16033			9·83967	
17·0	9·85900			10·14100	10·01946			9·98054	10·16046			9·83954	
18·0	9·85912	·6	7	10·14088	10·01971	·6	15	9·98029	10·16060	·6	8	9·83940	
19·0	9·85924			10·14076	10·01997			9·98003	10·16073			9·83927	
20·0	9·85936			10·14064	10·02022			9·97978	10·16086			9·83914	40′
21·0	9·85948	·7	8	10·14052	10·02047	·7	18	9·97953	10·16099	·7	9	9·83901	
22·0	9·85960			10·14040	10·02073			9·97927	10·16113			9·83888	
23·0	9·85972			10·14028	10·02098			9·97902	10·16126			9·83874	
24·0	9·85984	·8	10	10·14016	10·02123	·8	20	9·97877	10·16139	·8	11	9·83861	
25·0	9·85996			10·14004	10·02149			9·97852	10·16152			9·83848	35′
26·0	9·86008			10·13992	10·02174			9·97826	10·16166			9·83834	
27·0	9·86020	·9	11	10·13980	10·02199	·9	23	9·97801	10·16179	·9	12	9·83821	
28·0	9·86032			10·13968	10·02224			9·97776	10·16192			9·83808	
29·0	9·86044			10·13956	10·02250			9·97750	10·16206			9·83795	
30·0	9·86056			10·13944	10·02275			9·97725	10·16219			9·83781	30′
31·0	9·86068			10·13932	10·02300			9·97700	10·16232			9·83768	
32·0	9·86080	·1	1	10·13920	10·02326	·1	3	9·97674	10·16245	·1	1	9·83755	
33·0	9·86092			10·13908	10·02351			9·97649	10·16259			9·83741	
34·0	9·86104			10·13896	10·02376			9·97624	10·16272			9·83728	
35·0	9·86116	·2	2	10·13884	10·02402	·2	5	9·97599	10·16285	·2	3	9·83715	25′
36·0	9·86128			10·13872	10·02427			9·97573	10·16299			9·83701	
37·0	9·86140			10·13860	10·02452			9·97548	10·16312			9·83688	
38·0	9·86152	·3	4	10·13848	10·02477	·3	8	9·97523	10·16326	·3	4	9·83675	
39·0	9·86164			10·13836	10·02503			9·97497	10·16339			9·83661	
40·0	9·86176			10·13824	10·02528			9·97472	10·16352			9·83648	20′
41·0	9·86188			10·13812	10·02553			9·97447	10·16366			9·83634	
42·0	9·86200	·4	5	10·13800	10·02579	·4	10	9·97421	10·16379	·4	5	9·83621	
43·0	9·86212			10·13789	10·02604			9·97396	10·16393			9·83608	
44·0	9·86223			10·13777	10·02629			9·97371	10·16406			9·83594	
45·0	9·86235	·5	6	10·13765	10·02655	·5	13	9·97345	10·16419	·5	7	9·83581	15′
46·0	9·86247			10·13753	10·02680			9·97320	10·16433			9·83567	
47·0	9·86259			10·13741	10·02705			9·97295	10·16446			9·83554	
48·0	9·86271	·6	7	10·13729	10·02731	·6	15	9·97270	10·16460	·6	8	9·83540	
49·0	9·86283			10·13717	10·02756			9·97244	10·16473			9·83527	
50·0	9·86295			10·13705	10·02781			9·97219	10·16487			9·83513	10′
51·0	9·86306	·7	8	10·13694	10·02807	·7	18	9·97194	10·16500	·7	9	9·83500	
52·0	9·86318			10·13682	10·02832			9·97168	10·16514			9·83487	
53·0	9·86330			10·13670	10·02857			9·97143	10·16527			9·83473	
54·0	9·86342	·8	10	10·13658	10·02883	·8	20	9·97118	10·16541	·8	11	9·83460	
55·0	9·86354			10·13646	10·02908			9·97092	10·16554			9·83446	5′
56·0	9·86366			10·13634	10·02933			9·97067	10·16568			9·83433	
57·0	9·86377	·9	11	10·13623	10·02958	·9	23	9·97042	10·16581	·9	12	9·83419	
58·0	9·86389			10·13611	10·02984			9·97016	10·16595			9·83405	
59·0	9·86401			10·13599	10·03009			9·96991	10·16608			9·83392	
60·0	9·86413			10·13587	10·03034			9·96966	10·16622			9·83378	0′

′	·0 Log. 8·/9·	·0 Nat. 0·	·2 Log. 8·/9·	·2 Nat. 0·	·4 Log. 8·/9·	·4 Nat. 0·	·6 Log. 8·/9·	·6 Nat. 0·	·8 Log. 8·/9·	·8 Nat. 0·	Log. 8·/9·	Nat. 0·	′
00	97997	09549	98004	09551	98012	09553	98020	09554	98028	09556	98035	09558	59
01	98035	09558	98043	09559	98051	09561	98059	09563	98066	09565	98074	09566	58
02	98074	09566	98082	09568	98090	09570	98097	09571	98105	09573	98113	09575	57
03	98113	09575	98121	09577	98129	09578	98136	09580	98144	09582	98152	09583	56
04	98152	09583	98160	09585	98167	09587	98175	09589	98183	09590	98191	09592	55
05	98191	09592	98198	09594	98206	09595	98214	09597	98222	09599	98229	09601	54
06	98229	09601	98237	09602	98245	09604	98253	09606	98260	09607	98268	09609	53
07	98268	09609	98276	09611	98284	09613	98291	09614	98299	09616	98307	09618	52
08	98307	09618	98315	09619	98322	09621	98330	09623	98338	09625	98346	09626	51
09	98346	09626	98353	09628	98361	09630	98369	09631	98377	09633	98384	09635	50
10	98384	09635	98392	09637	98400	09638	98408	09640	98415	09642	98423	09643	49
11	98423	09643	98431	09645	98438	09647	98446	09649	98454	09650	98462	09652	48
12	98462	09652	98469	09654	98477	09655	98485	09657	98493	09659	98500	09661	47
13	98500	09661	98508	09662	98516	09664	98524	09666	98531	09667	98539	09669	46
14	98539	09669	98547	09671	98554	09673	98562	09674	98570	09676	98578	09678	45
15	98578	09678	98585	09679	98593	09681	98601	09683	98608	09685	98616	09686	44
16	98616	09686	98624	09688	98632	09690	98639	09692	98647	09693	98655	09695	43
17	98655	09695	98662	09697	98670	09698	98678	09700	98685	09702	98693	09704	42
18	98693	09704	98701	09705	98709	09707	98716	09709	98724	09710	98732	09712	41
19	98732	09712	98739	09714	98747	09716	98755	09717	98763	09719	98770	09721	40
20	98770	09721	98778	09723	98786	09724	98793	09726	98801	09728	98809	09729	39
21	98809	09729	98816	09731	98824	09733	98832	09735	98840	09736	98847	09738	38
22	98847	09738	98855	09740	98863	09742	98870	09743	98878	09745	98886	09747	37
23	98886	09747	98893	09748	98901	09750	98909	09752	98916	09754	98924	09755	36
24	98924	09755	98932	09757	98940	09759	98947	09760	98955	09762	98963	09764	35
25	98963	09764	98970	09766	98978	09767	98986	09769	98993	09771	99001	09773	34
26	99001	09773	99009	09774	99016	09776	99024	09778	99032	09779	99039	09781	33
27	99039	09781	99047	09783	99055	09785	99062	09786	99070	09788	99078	09790	32
28	99078	09790	99085	09792	99093	09793	99101	09795	99108	09797	99116	09799	31
29	99116	09799	99124	09800	99131	09802	99139	09804	99147	09805	99154	09807	30
30	99154	09807	99162	09809	99170	09811	99177	09812	99185	09814	99193	09816	29
31	99193	09816	99200	09818	99208	09819	99216	09821	99223	09823	99231	09824	28
32	99231	09824	99238	09826	99246	09828	99254	09830	99262	09831	99269	09833	27
33	99269	09833	99277	09835	99284	09837	99292	09838	99300	09840	99307	09842	26
34	99307	09842	99315	09844	99323	09845	99330	09847	99338	09849	99346	09850	25
35	99346	09850	99353	09852	99361	09854	99369	09856	99376	09857	99384	09859	24
36	99384	09859	99391	09861	99399	09863	99407	09864	99414	09866	99422	09868	23
37	99422	09868	99430	09870	99437	09871	99445	09873	99453	09875	99460	09877	22
38	99460	09877	99468	09878	99475	09880	99483	09882	99491	09883	99498	09885	21
39	99498	09885	99506	09887	99514	09889	99521	09890	99529	09892	99536	09894	20
40	99536	09894	99544	09896	99552	09897	99559	09899	99567	09901	99575	09903	19
41	99575	09903	99582	09904	99590	09906	99598	09908	99605	09909	99613	09911	18
42	99613	09911	99620	09913	99628	09915	99636	09916	99643	09918	99651	09920	17
43	99651	09920	99658	09922	99666	09923	99674	09925	99681	09927	99689	09929	16
44	99689	09929	99696	09930	99704	09932	99712	09934	99719	09936	99727	09937	15
45	99727	09937	99734	09939	99742	09941	99750	09943	99757	09944	99765	09946	14
46	99765	09946	99773	09948	99780	09949	99788	09951	99795	09953	99803	09955	13
47	99803	09955	99811	09956	99818	09958	99826	09960	99833	09962	99841	09963	12
48	99841	09963	99848	09965	99856	09967	99864	09969	99871	09970	99879	09972	11
49	99879	09972	99886	09974	99894	09975	99902	09977	99909	09979	99917	09981	10
50	99917	09981	99924	09983	99932	09984	99940	09986	99947	09988	99955	09990	09
51	99955	09990	99962	09991	99970	09993	99977	09995	99985	09997	99993	09998	08
52	99993	09998	00000	10000	00008	10002	00015	10004	00023	10005	00031	10007	07
53	00031	10007	00038	10009	00046	10011	00053	10012	00061	10014	00068	10016	06
54	00068	10016	00076	10018	00084	10019	00091	10021	00099	10023	00106	10025	05
55	00106	10025	00114	10026	00121	10028	00129	10030	00137	10032	00144	10033	04
56	00144	10033	00152	10035	00159	10037	00167	10039	00174	10040	00182	10042	03
57	00182	10042	00190	10044	00197	10046	00205	10047	00212	10049	00220	10051	02
58	00220	10051	00227	10052	00235	10054	00242	10056	00250	10058	00258	10059	01
59	00258	10059	00265	10061	00273	10063	00280	10065	00288	10066	00295	10068	00
	·8		·6		·4		·2		·0				

PARTS for 0′·1 :—   LOGS. 4 ;   NATURALS, 1.

323°

HAVERSINES

	·0		·2		·4		·6		·8		·8		
	Log.	Nat.	Log.	Nat.	Log.	Nat.	Log.	Nat.	Log.	Nat.	Log.	Nat.	
′	9·	0·	9·	0·	9·	0·	9·	0·	9·	0·	9·	0·	′
00	21863	16543	21868	16546	21874	16548	21880	16550	21885	16552	21891	16554	59
01	21891	16554	21897	16556	21902	16559	21908	16561	21914	16563	21919	16565	58
02	21919	16565	21925	16567	21931	16569	21936	16572	21942	16574	21948	16576	57
03	21948	16576	21953	16578	21959	16580	21965	16582	21970	16585	21976	16587	56
04	21976	16587	21982	16589	21987	16591	21993	16593	21999	16595	22004	16598	55
05	22004	16598	22010	16600	22016	16602	22021	16604	22027	16606	22033	16608	54
06	22033	16608	22038	16611	22044	16613	22050	16615	22055	16617	22061	16619	53
07	22061	16619	22067	16621	22072	16624	22078	16626	22084	16628	22089	16630	52
08	22089	16630	22095	16632	22101	16634	22106	16637	22112	16639	22118	16641	51
09	22118	16641	22123	16643	22129	16645	22135	16647	22140	16650	22146	16652	50
10	22146	16652	22151	16654	22157	16656	22163	16658	22169	16660	22174	16663	49
11	22174	16663	22180	16665	22185	16667	22191	16669	22197	16671	22202	16673	48
12	22202	16673	22208	16676	22214	16678	22219	16680	22225	16682	22231	16684	47
13	22231	16684	22236	16686	22242	16689	22248	16691	22253	16693	22259	16695	46
14	22259	16695	22264	16697	22270	16699	22276	16702	22281	16704	22287	16706	45
15	22287	16706	22293	16708	22298	16710	22304	16712	22310	16715	22315	16717	44
16	22315	16717	22321	16719	22326	16721	22332	16723	22338	16725	22343	16728	43
17	22343	16728	22349	16730	22355	16732	22360	16734	22366	16736	22372	16738	42
18	22372	16738	22377	16741	22383	16743	22389	16745	22394	16747	22400	16749	41
19	22400	16749	22405	16752	22411	16754	22417	16756	22422	16758	22428	16760	40
20	22428	16760	22434	16762	22439	16765	22445	16767	22450	16769	22456	16771	39
21	22456	16771	22462	16773	22467	16775	22473	16778	22479	16780	22484	16782	38
22	22484	16782	22490	16784	22495	16786	22501	16788	22507	16791	22512	16793	37
23	22512	16793	22518	16795	22524	16797	22529	16799	22535	16802	22540	16804	36
24	22540	16804	22546	16806	22552	16808	22557	16810	22563	16812	22569	16815	35
25	22569	16815	22574	16817	22580	16819	22585	16821	22591	16823	22597	16825	34
26	22597	16825	22602	16828	22608	16830	22614	16832	22619	16834	22625	16836	33
27	22625	16836	22630	16839	22636	16841	22642	16843	22647	16845	22653	16847	32
28	22653	16847	22658	16849	22664	16852	22670	16854	22675	16856	22681	16858	31
29	22681	16858	22686	16860	22692	16862	22698	16865	22703	16867	22709	16869	30
30	22709	16869	22715	16871	22720	16873	22726	16876	22731	16878	22737	16880	29
31	22737	16880	22743	16882	22748	16884	22754	16886	22759	16889	22765	16891	28
32	22765	16891	22771	16893	22776	16895	22782	16897	22788	16900	22793	16902	27
33	22793	16902	22799	16904	22804	16906	22810	16908	22815	16910	22821	16913	26
34	22821	16913	22827	16915	22832	16917	22838	16919	22843	16921	22849	16924	25
35	22849	16924	22854	16926	22860	16928	22866	16930	22871	16932	22877	16934	24
36	22877	16934	22883	16937	22888	16939	22894	16941	22899	16943	22905	16945	23
37	22905	16945	22911	16947	22916	16950	22922	16952	22927	16954	22933	16956	22
38	22933	16956	22939	16958	22944	16961	22950	16963	22955	16965	22961	16967	21
39	22961	16967	22966	16969	22972	16972	22978	16974	22983	16976	22989	16978	20
40	22989	16978	22994	16980	23000	16982	23006	16985	23011	16987	23017	16989	19
41	23017	16989	23022	16991	23028	16993	23033	16996	23039	16998	23045	17000	18
42	23045	17000	23050	17002	23056	17004	23061	17006	23067	17009	23073	17011	17
43	23073	17011	23078	17013	23084	17015	23089	17017	23095	17020	23100	17022	16
44	23100	17022	23106	17024	23112	17026	23117	17028	23123	17030	23128	17033	15
45	23128	17033	23134	17035	23140	17037	23145	17039	23151	17041	23156	17044	14
46	23156	17044	23162	17046	23167	17048	23173	17050	23179	17052	23184	17055	13
47	23184	17055	23190	17057	23195	17059	23201	17061	23206	17063	23212	17066	12
48	23212	17066	23218	17068	23223	17070	23229	17072	23234	17074	23240	17076	11
49	23240	17076	23245	17079	23251	17081	23256	17083	23262	17085	23268	17087	10
50	23268	17087	23273	17090	23279	17092	23284	17094	23290	17096	23295	17098	09
51	23295	17098	23301	17101	23307	17103	23312	17105	23318	17107	23323	17109	08
52	23323	17109	23329	17112	23334	17114	23340	17116	23345	17118	23351	17120	07
53	23351	17120	23357	17122	23362	17125	23368	17127	23373	17129	23379	17131	06
54	23379	17131	23384	17133	23390	17136	23395	17138	23401	17140	23407	17142	05
55	23407	17142	23412	17144	23418	17147	23423	17149	23429	17151	23434	17153	04
56	23434	17153	23440	17155	23446	17158	23451	17160	23457	17162	23462	17164	03
57	23462	17164	23468	17166	23473	17169	23479	17171	23484	17173	23490	17175	02
58	23490	17175	23496	17177	23501	17179	23507	17182	23512	17184	23518	17186	01
59	23518	17186	23523	17188	23529	17190	23534	17193	23540	17195	23545	17197	00
	·8		·6		·4		·2		·0				

PARTS for 0′·1 :— LOGS. 3; NATURALS, 1.

311°

HAVERSINES

′	·0 Log.	·0 Nat.	·2 Log.	·2 Nat.	·4 Log.	·4 Nat.	·6 Log.	·6 Nat.	·8 Log.	·8 Nat.	Log.	Nat.	′
	9·	0·	9·	0·	9·	0·	9·	0·	9·	0·	9·	0·	
00	39794	25000	39798	25003	39803	25005	39807	25008	39811	25010	39816	25013	59
01	39816	25013	39820	25015	39825	25018	39829	25020	39833	25023	39838	25025	58
02	39838	25025	39842	25028	39847	25030	39851	25033	39855	25035	39860	25038	57
03	39860	25038	39864	25040	39868	25043	39873	25045	39877	25048	39881	25050	56
04	39881	25050	39886	25053	39890	25055	39895	25058	39899	25060	39903	25063	55
05	39903	25063	39908	25066	39912	25068	39916	25071	39921	25073	39925	25076	54
06	39925	25076	39930	25078	39934	25081	39938	25083	39943	25086	39947	25088	53
07	39947	25088	39951	25091	39956	25093	39960	25096	39964	25098	39969	25101	52
08	39969	25101	39973	25103	39978	25106	39982	25108	39986	25111	39991	25113	51
09	39991	25113	39995	25116	39999	25118	40004	25121	40008	25124	40012	25126	50
10	40012	25126	40017	25129	40021	25131	40025	25134	40030	25136	40034	25139	49
11	40034	25139	40039	25141	40043	25144	40047	25146	40052	25149	40056	25151	48
12	40056	25151	40060	25154	40065	25156	40069	25159	40073	25161	40078	25164	47
13	40078	25164	40082	25166	40087	25169	40091	25171	40095	25174	40100	25177	46
14	40100	25177	40104	25179	40108	25182	40113	25184	40117	25187	40121	25189	45
15	40121	25189	40126	25192	40130	25194	40134	25197	40139	25199	40143	25202	44
16	40143	25202	40148	25204	40152	25207	40156	25209	40161	25212	40165	25214	43
17	40165	25214	40169	25217	40174	25219	40178	25222	40182	25225	40187	25227	42
18	40187	25227	40191	25230	40195	25232	40200	25235	40204	25237	40208	25240	41
19	40208	25240	40213	25242	40217	25245	40221	25247	40226	25250	40230	25252	40
20	40230	25252	40235	25255	40239	25257	40243	25260	40247	25262	40252	25265	39
21	40252	25265	40256	25268	40261	25270	40265	25273	40269	25275	40274	25278	38
22	40274	25278	40278	25280	40282	25283	40287	25285	40291	25288	40295	25290	37
23	40295	25290	40300	25293	40304	25295	40308	25298	40313	25300	40317	25303	36
24	40317	25303	40321	25305	40326	25308	40330	25310	40334	25313	40339	25316	35
25	40339	25316	40343	25318	40347	25321	40352	25323	40356	25326	40360	25328	34
26	40360	25328	40365	25331	40369	25333	40373	25336	40378	25338	40382	25341	33
27	40382	25341	40387	25343	40391	25346	40395	25348	40399	25351	40404	25354	32
28	40404	25354	40408	25356	40412	25359	40417	25361	40421	25364	40425	25366	31
29	40425	25366	40430	25369	40434	25371	40438	25374	40443	25376	40447	25379	30
30	40447	25379	40452	25381	40456	25384	40460	25386	40464	25389	40469	25391	29
31	40469	25391	40473	25394	40477	25397	40482	25399	40486	25402	40490	25404	28
32	40490	25404	40495	25407	40499	25409	40503	25412	40508	25414	40512	25417	27
33	40512	25417	40517	25419	40521	25422	40525	25424	40529	25427	40534	25429	26
34	40534	25429	40538	25432	40542	25435	40547	25437	40551	25440	40555	25442	25
35	40555	25442	40560	25445	40564	25447	40568	25450	40573	25452	40577	25455	24
36	40577	25455	40581	25457	40586	25460	40590	25462	40594	25465	40599	25467	23
37	40599	25467	40603	25470	40607	25473	40612	25475	40616	25478	40620	25480	22
38	40620	25480	40625	25483	40629	25485	40633	25488	40637	25490	40642	25493	21
39	40642	25493	40646	25495	40650	25498	40655	25500	40659	25503	40663	25506	20
40	40663	25506	40668	25508	40672	25511	40676	25513	40681	25516	40685	25519	19
41	40685	25518	40689	25521	40694	25523	40698	25526	40702	25528	40707	25531	18
42	40707	25531	40711	25533	40715	25536	40720	25538	40724	25541	40728	25544	17
43	40728	25544	40733	25546	40737	25549	40741	25551	40745	25554	40750	25556	16
44	40750	25556	40754	25559	40758	25561	40763	25564	40767	25566	40771	25569	15
45	40771	25569	40776	25571	40780	25574	40784	25577	40788	25579	40793	25582	14
46	40793	25582	40797	25584	40801	25587	40806	25589	40810	25592	40814	25594	13
47	40814	25594	40819	25597	40823	25599	40827	25602	40832	25604	40836	25607	12
48	40836	25607	40840	25610	40844	25612	40849	25615	40853	25617	40858	25620	11
49	40858	25620	40862	25622	40866	25625	40870	25627	40875	25630	40879	25632	10
50	40879	25632	40883	25635	40888	25637	40892	25640	40896	25643	40900	25645	09
51	40900	25645	40905	25648	40909	25650	40913	25653	40918	25655	40922	25658	08
52	40922	25658	40926	25660	40931	25663	40935	25665	40939	25668	40943	25671	07
53	40943	25671	40948	25673	40952	25676	40956	25678	40961	25681	40965	25683	06
54	40965	25683	40969	25686	40974	25688	40978	25691	40982	25693	40986	25696	05
55	40986	25696	40991	25698	40995	25701	40999	25704	41004	25706	41008	25709	04
56	41008	25709	41012	25711	41017	25714	41021	25716	41025	25719	41029	25721	03
57	41029	25721	41034	25724	41038	25726	41042	25729	41047	25732	41051	25734	02
58	41051	25734	41055	25737	41059	25739	41064	25742	41068	25744	41072	25747	01
59	41072	25747	41077	25749	41081	25752	41085	25754	41090	25757	41094	25760	00

| | ·8 | | ·6 | | ·4 | | ·2 | | ·0 | | | | |

299°

PARTS for 0′·1 :—   LOGS. 2 ;   NATURALS, 1.

HAVERSINES

		·0		·2		·4		·6		·8		
		Log.	Nat.	Log.	Nat.	Log.	Nat.	Log.	Nat.	Log.	Nat.	
	′	9·	0·	9·	0·	9·	0·	9·	0·	9·	0·	′
00		42368	26526	42372	26529	42376	26532	42380	26534	42385	26537	42389	26539	59
01		42389	26539	42393	26542	42397	26544	42402	26547	42406	26550	42410	26552	58
02		42410	26552	42414	26555	42418	26557	42423	26560	42427	26562	42431	26565	57
03		42431	26565	42435	26568	42439	26570	42444	26573	42448	26575	42452	26578	56
04		42452	26578	42456	26580	42460	26583	42465	26586	42469	26588	42473	26591	55
05		42473	26591	42477	26593	42481	26596	42485	26598	42490	26601	42494	26604	54
06		42494	26604	42498	26606	42502	26609	42506	26611	42511	26614	42515	26616	53
07		42515	26616	42519	26619	42523	26621	42527	26624	42532	26627	42536	26629	52
08		42536	26629	42540	26632	42544	26634	42548	26637	42553	26640	42557	26642	51
09		42557	26642	42561	26645	42565	26647	42569	26650	42574	26652	42578	26655	50
10		42578	26655	42582	26658	42586	26660	42590	26663	42595	26665	42599	26668	49
11		42599	26668	42603	26670	42607	26673	42611	26676	42615	26678	42620	26681	48
12		42620	26681	42624	26683	42628	26686	42632	26688	42636	26691	42641	26694	47
13		42641	26694	42645	26696	42649	26699	42653	26701	42657	26704	42662	26706	46
14		42662	26706	42666	26709	42670	26712	42674	26714	42678	26717	42682	26719	45
15		42682	26719	42687	26722	42691	26724	42695	26727	42699	26730	42703	26732	44
16		42703	26732	42708	26735	42712	26737	42716	26740	42720	26742	42724	26745	43
17		42724	26745	42728	26748	42733	26750	42737	26753	42741	26755	42745	26758	42
18		42745	26758	42749	26760	42754	26763	42758	26766	42762	26768	42766	26771	41
19		42766	26771	42770	26773	42775	26776	42779	26779	42783	26781	42787	26784	40
20		42787	26784	42791	26786	42795	26789	42799	26791	42804	26794	42808	26797	39
21		42808	26797	42812	26799	42816	26802	42820	26804	42825	26807	42829	26809	38
22		42829	26809	42833	26812	42837	26815	42841	26817	42845	26820	42850	26822	37
23		42850	26822	42854	26825	42858	26827	42862	26830	42866	26833	42870	26835	36
24		42870	26835	42875	26838	42879	26840	42883	26843	42887	26846	42891	26848	35
25		42891	26848	42896	26851	42900	26853	42904	26856	42908	26858	42912	26861	34
26		42912	26861	42916	26864	42921	26866	42925	26869	42929	26871	42933	26874	33
27		42933	26874	42937	26876	42941	26879	42945	26882	42950	26884	42954	26887	32
28		42954	26887	42958	26889	42962	26892	42966	26894	42970	26897	42975	26900	31
29		42975	26900	42979	26902	42983	26905	42987	26907	42991	26910	42996	26913	30
30		42996	26913	43000	26915	43004	26918	43008	26920	43012	26923	43016	26925	29
31		43016	26925	43021	26928	43025	26931	43029	26933	43033	26936	43037	26938	28
32		43037	26938	43041	26941	43046	26944	43050	26946	43054	26949	43058	26951	27
33		43058	26951	43062	26954	43066	26956	43070	26959	43075	26962	43079	26964	26
34		43079	26964	43083	26967	43087	26969	43091	26972	43095	26975	43100	26977	25
35		43100	26977	43104	26980	43108	26982	43112	26985	43116	26987	43120	26990	24
36		43120	26990	43125	26993	43129	26995	43133	26998	43137	27000	13141	27003	23
37		43141	27003	43145	27005	43149	27008	43153	27011	43158	27013	43162	27016	22
38		43162	27016	43166	27018	43170	27021	43174	27024	43178	27026	43183	27029	21
39		43183	27029	43187	27031	43191	27034	43195	27037	43199	27039	43203	27042	20
40		43203	27042	43208	27044	43212	27047	43216	27049	43220	27052	43224	27055	19
41		43224	27055	43228	27057	43232	27060	43237	27062	43241	27065	43245	27068	18
42		43245	27068	43249	27070	43253	27073	43257	27075	43261	27078	43266	27080	17
43		43266	27080	43270	27083	43274	27086	43278	27088	43282	27091	43286	27093	16
44		43286	27093	43291	27096	43295	27099	43299	27101	43303	27104	43307	27106	15
45		43307	27106	43311	27109	43315	27111	43319	27114	43324	27117	43328	27119	14
46		43328	27119	43332	27122	43336	27124	43340	27127	43344	27130	43348	27132	13
47		43348	27132	43353	27135	43357	27137	43361	27140	43365	27143	43369	27145	12
48		43369	27145	43373	27148	43377	27150	43382	27153	43386	27155	43390	27158	11
49		43390	27158	43394	27161	43398	27163	43402	27166	43406	27168	43411	27171	10
50		43411	27171	43415	27174	43419	27176	43423	27179	43427	27181	43431	27184	09
51		43431	27184	43435	27186	43439	27189	43444	27192	43448	27194	43452	27197	08
52		43452	27197	43456	27199	43460	27202	43464	27205	43468	27207	43473	27210	07
53		43473	27210	43477	27212	43481	27215	43485	27218	43489	27220	43493	27223	06
54		43493	27223	43497	27225	43501	27228	43506	27231	43510	27233	43514	27236	05
55		43514	27236	43518	27238	43522	27241	43526	27243	43530	27246	43535	27249	04
56		43535	27249	43539	27251	43543	27254	43547	27256	43551	27259	43555	27262	03
57		43555	27262	43559	27264	43563	27267	43568	27269	43572	27272	43576	27275	02
58		43576	27275	43580	27277	43584	27280	43588	27282	43592	27285	43596	27288	01
59		43596	27288	43601	27290	43605	27293	43609	27295	43613	27298	43617	27300	00
		·8		·6		·4		·2		·0	←	↑	

PARTS for 0′·1 :—   LOGS. 2 ;   NATURALS, 1.

Lat. °	45° 315°	46° 314°	47° 313°	48° 312°	49° 311°	50° 310°	51° 309°	52° 308°	53° 307°	54° 306°	55° 305°	56° 304°	57° 303°	58° 302°	59° 301°	60° 300°	Lat. °
0	·00	·00	·00	·00	·00	·00	·00	·00	·00	·00	·00	·00	·00	·00	·00	·00	0
1	·02	·02	·02	·02	·02	·01	·01	·01	·01	·01	·01	·01	·01	·01	·01	·01	1
2	·03	·03	·03	·03	·03	·03	·03	·03	·03	·03	·02	·02	·02	·02	·02	·02	2
3	·05	·05	·05	·05	·05	·04	·04	·04	·04	·04	·04	·04	·03	·03	·03	·03	3
4	·07	·07	·07	·06	·06	·06	·06	·05	·05	·05	·05	·05	·05	·04	·04	·04	4
5	·09	·08	·08	·08	·08	·07	·07	·07	·07	·06	·06	·06	·06	·05	·05	·05	5
6	·11	·10	·10	·09	·09	·09	·09	·08	·08	·08	·07	·07	·07	·07	·06	·06	6
7	·12	·12	·11	·11	·11	·10	·10	·10	·09	·09	·09	·08	·08	·08	·07	·07	7
8	·14	·14	·13	·13	·12	·12	·11	·11	·11	·10	·10	·09	·09	·09	·08	·08	8
9	·16	·15	·15	·14	·14	·13	·13	·12	·12	·12	·11	·11	·10	·10	·10	·09	9
10	·18	·17	·16	·16	·15	·15	·14	·14	·13	·13	·12	·12	·11	·11	·11	·10	10
11	·19	·19	·18	·18	·17	·16	·16	·15	·15	·14	·14	·13	·13	·12	·12	·11	11
12	·21	·21	·20	·19	·18	·18	·17	·17	·16	·15	·15	·14	·14	·13	·13	·12	12
13	·23	·22	·22	·21	·20	·19	·19	·18	·17	·17	·16	·16	·15	·14	·14	·13	13
14	·25	·24	·23	·22	·22	·21	·20	·19	·19	·18	·17	·17	·16	·16	·15	·14	14
15	·27	·26	·25	·24	·23	·22	·22	·21	·20	·19	·19	·18	·17	·17	·16	·15	15
16	·29	·28	·27	·26	·25	·24	·23	·22	·22	·21	·20	·19	·19	·18	·17	·17	16
17	·31	·30	·29	·28	·27	·26	·25	·24	·23	·22	·21	·21	·20	·19	·18	·18	17
18	·32	·31	·30	·29	·28	·27	·26	·25	·24	·24	·23	·22	·21	·20	·20	·19	18
19	·34	·33	·32	·31	·30	·29	·28	·27	·26	·25	·24	·23	·22	·22	·21	·20	19
20	·36	·35	·34	·33	·32	·31	·29	·28	·27	·26	·25	·25	·24	·23	·22	·21	20
21	·38	·37	·36	·35	·33	·32	·31	·30	·29	·28	·27	·26	·25	·24	·23	·22	21
22	·40	·39	·38	·36	·35	·34	·33	·32	·30	·29	·28	·27	·26	·25	·24	·23	22
23	·42	·41	·40	·38	·37	·36	·34	·33	·32	·31	·30	·29	·28	·27	·26	·25	23
24	·45	·43	·42	·40	·39	·37	·36	·35	·34	·32	·31	·30	·29	·28	·27	·26	24
25	·47	·45	·44	·42	·41	·39	·38	·36	·35	·34	·33	·31	·30	·29	·28	·27	25
26	·49	·47	·46	·44	·42	·41	·39	·38	·37	·35	·34	·33	·32	·30	·29	·28	26
27	·51	·49	·48	·46	·44	·43	·41	·40	·38	·37	·36	·34	·33	·32	·31	·29	27
28	·53	·51	·50	·48	·46	·45	·43	·42	·40	·39	·37	·36	·35	·33	·32	·31	28
29	·55	·54	·52	·50	·48	·47	·45	·43	·42	·40	·39	·37	·36	·35	·33	·32	29
30	·58	·56	·54	·52	·50	·48	·47	·45	·44	·42	·40	·39	·37	·36	·35	·33	30
31	·60	·58	·56	·54	·52	·50	·49	·47	·45	·44	·42	·40	·39	·38	·36	·35	31
32	·62	·60	·58	·56	·54	·52	·51	·49	·47	·45	·44	·42	·41	·39	·38	·36	32
33	·65	·63	·61	·58	·56	·55	·53	·51	·49	·47	·45	·44	·42	·41	·39	·37	33
34	·67	·65	·63	·61	·59	·57	·55	·53	·51	·49	·47	·46	·44	·42	·41	·39	34
35	·70	·68	·65	·63	·61	·59	·57	·55	·53	·51	·49	·47	·45	·44	·42	·40	35
36	·73	·70	·68	·65	·63	·61	·59	·57	·55	·53	·51	·49	·47	·45	·44	·42	36
37	·75	·73	·70	·68	·66	·63	·61	·59	·57	·55	·53	·51	·49	·47	·45	·44	37
38	·78	·75	·73	·70	·68	·66	·63	·61	·59	·57	·55	·53	·51	·49	·47	·45	38
39	·81	·78	·76	·73	·70	·68	·66	·63	·61	·59	·57	·55	·53	·51	·49	·47	39
40	·84	·81	·78	·76	·73	·70	·68	·66	·63	·61	·59	·57	·55	·52	·50	·48	40
41	·87	·84	·81	·78	·76	·73	·70	·68	·66	·63	·61	·59	·56	·54	·52	·50	41
42	·90	·87	·84	·81	·78	·76	·73	·70	·68	·65	·63	·61	·58	·56	·54	·52	42
43	·93	·90	·87	·84	·81	·78	·76	·73	·70	·68	·65	·63	·61	·58	·56	·54	43
44	·97	·93	·90	·87	·84	·81	·78	·75	·73	·70	·68	·65	·63	·60	·58	·56	44
45	1·00	·97	·93	·90	·87	·84	·81	·78	·75	·73	·70	·68	·65	·63	·60	·58	45
46	1·04	1·00	·97	·93	·90	·87	·84	·81	·78	·75	·73	·70	·67	·65	·62	·60	46
47	1·07	1·04	1·00	·97	·93	·90	·87	·84	·81	·78	·75	·72	·70	·67	·64	·62	47
48	1·11	1·07	1·04	1·00	·97	·93	·90	·87	·84	·81	·78	·75	·72	·69	·67	·64	48
49	1·15	1·11	1·07	1·04	1·00	·97	·93	·90	·87	·84	·81	·78	·75	·72	·69	·66	49
50	1·19	1 15	1·11	1·07	1·04	1·00	·97	·93	·90	·87	·83	·80	·77	·75	·72	·69	50
51	1·23	1·19	1·15	1·11	1·07	1·04	1·00	·97	·93	·90	·86	·83	·80	·77	·74	·71	51
52	1·28	1·24	1·19	1·15	1·11	1·07	1·04	1·00	·96	·93	·90	·86	·83	·80	·77	·74	52
53	1·33	1·28	1·24	1·19	1·15	1·11	1·07	1·04	1·00	·96	·93	·90	·86	·83	·80	·77	53
54	1·38	1·33	1·28	1·24	1·20	1·15	1·11	1·08	1·04	1·00	·96	·93	·89	·86	·83	·79	54
55	1·43	1·38	1·33	1·29	1·24	1·20	1·16	1·12	1·08	1·04	1·00	·96	·93	·89	·86	·82	55
56	1·48	1·43	1·38	1·34	1·29	1·24	1·20	1·16	1·12	1·08	1·04	1·00	·96	·93	·89	·86	56
57	1·54	1·49	1·44	1·39	1·34	1·29	1·25	1·20	1·16	1·12	1·08	1·04	1·00	·96	·93	·89	57
58	1·60	1·55	1·49	1·44	1·39	1·34	1·30	1·25	1·21	1·16	1·12	1·08	1·04	1·00	·96	·92	58
59	1·66	1·61	1·55	1·50	1·45	1·40	1·35	1·30	1·25	1·21	1·17	1·12	1·08	1·04	1·00	·96	59
60	1·73	1·67	1·62	1·56	1·51	1·45	1·40	1·35	1·31	1·26	1·21	1·17	1·12	1·08	1·04	1·00	60
Lat.	135° 225°	134° 226°	133° 227°	132° 228°	131° 229°	130° 230°	129° 231°	128° 232°	127° 233°	126° 234°	125° 235°	124° 236°	123° 237°	122° 238°	121° 239°	120° 240°	Lat.

A — HOUR ANGLE — A

A—Named opposite to Latitude, **except** when Hour Angle is between 90° and 270°

A—Named opposite to Latitude, **except** when Hour Angle is between 90° and 270°

Dec.°	45° 315°	46° 314°	47° 313°	48° 312°	49° 311°	50° 310°	51° 309°	52° 308°	53° 307°	54° 306°	55° 305°	56° 304°	57° 303°	58° 302°	59° 301°	60° 300°	Dec.°
0	·00	·00	·00	·00	·00	·00	·00	·00	·00	·00	·00	·00	·00	·00	·00	·00	0
1	·02	·02	·02	·02	·02	·02	·02	·02	·02	·02	·02	·02	·02	·02	·02	·02	1
2	·05	·05	·05	·05	·05	·05	·04	·04	·04	·04	·04	·04	·04	·04	·04	·04	2
3	·07	·07	·07	·07	·07	·07	·07	·07	·07	·06	·06	·06	·06	·06	·06	·06	3
4	·10	·10	·10	·09	·09	·09	·09	·09	·09	·09	·09	·08	·08	·08	·08	·08	4
5	·12	·12	·12	·12	·12	·11	·11	·11	·11	·11	·11	·11	·10	·10	·10	·10	5
6	·15	·15	·14	·14	·14	·14	·14	·13	·13	·13	·13	·13	·13	·12	·12	·12	6
7	·17	·17	·17	·17	·16	·16	·16	·16	·15	·15	·15	·15	·15	·14	·14	·14	7
8	·20	·20	·19	·19	·18	·18	·18	·18	·18	·17	·17	·17	·17	·17	·16	·16	8
9	·22	·22	·22	·21	·21	·21	·20	·20	·20	·20	·19	·19	·19	·19	·18	·18	9
10	·25	·25	·24	·24	·23	·23	·23	·22	·22	·22	·22	·21	·21	·21	·21	·20	10
11	·27	·27	·27	·26	·26	·25	·25	·25	·24	·24	·24	·23	·23	·23	·23	·22	11
12	·30	·30	·29	·29	·28	·28	·27	·27	·27	·26	·26	·25	·25	·25	·25	·25	12
13	·33	·32	·32	·31	·31	·30	·30	·29	·29	·29	·28	·28	·28	·27	·27	·27	13
14	·35	·35	·34	·34	·33	·33	·32	·32	·31	·31	·30	·30	·30	·29	·29	·29	14
15	·38	·37	·37	·36	·36	·34	·34	·34	·34	·33	·33	·32	·32	·32	·31	·31	15
16	·41	·40	·39	·39	·38	·37	·37	·36	·36	·35	·35	·35	·34	·34	·33	·33	16
17	·43	·43	·42	·41	·41	·40	·39	·39	·38	·38	·37	·37	·36	·36	·36	·35	17
18	·46	·45	·44	·44	·43	·42	·42	·41	·41	·40	·40	·39	·39	·38	·38	·38	18
19	·49	·48	·47	·46	·46	·45	·44	·44	·43	·43	·42	·42	·41	·41	·40	·40	19
20	·51	·51	·50	·49	·48	·48	·47	·46	·46	·45	·44	·44	·43	·43	·42	·42	20
21	·54	·53	·52	·52	·51	·50	·49	·49	·48	·47	·47	·46	·46	·45	·45	·44	21
22	·57	·56	·55	·54	·54	·53	·52	·51	·51	·50	·49	·49	·48	·48	·47	·47	22
23	·60	·59	·58	·57	·56	·55	·55	·54	·53	·52	·52	·51	·51	·50	·50	·49	23
24	·63	·62	·61	·60	·59	·58	·57	·57	·56	·55	·54	·54	·53	·53	·52	·51	24
25	·66	·65	·64	·63	·62	·61	·60	·59	·58	·58	·57	·56	·56	·55	·54	·54	25
26	·69	·68	·67	·66	·65	·64	·63	·62	·61	·60	·60	·59	·58	·58	·57	·56	26
27	·72	·71	·70	·69	·68	·67	·66	·65	·64	·63	·62	·61	·61	·60	·59	·59	27
28	·75	·74	·73	·72	·70	·69	·68	·67	·67	·66	·65	·64	·63	·63	·62	·61	28
29	·78	·77	·76	·75	·73	·72	·71	·70	·69	·69	·68	·67	·66	·65	·65	·64	29
30	·82	·80	·79	·78	·76	·75	·74	·73	·72	·71	·70	·70	·69	·68	·67	·67	30
31	·85	·84	·82	·81	·80	·78	·77	·76	·75	·74	·73	·72	·72	·71	·70	·69	31
32	·88	·87	·85	·84	·83	·82	·80	·79	·78	·77	·76	·75	·75	·74	·73	·72	32
33	·92	·90	·89	·87	·85	·85	·84	·82	·81	·80	·79	·78	·77	·77	·76	·75	33
34	·96	·94	·92	·91	·89	·88	·87	·86	·84	·83	·82	·81	·80	·80	·79	·78	34
35	·99	·97	·96	·94	·93	·91	·90	·89	·88	·87	·85	·84	·83	·83	·82	·81	35
36	1·03	1·01	·99	·98	·96	·95	·93	·92	·91	·90	·89	·88	·87	·86	·85	·84	36
37	1·07	1·05	1·03	1·01	1·00	·98	·97	·96	·94	·93	·92	·91	·90	·89	·88	·87	37
38	1·11	1·09	1·07	1·05	1·04	1·02	1·00	·99	·98	·97	·95	·94	·93	·92	·91	·90	38
39	1·15	1·13	1·11	1·09	1·07	1·06	1·04	1·03	1·01	1·00	·99	·98	·97	·95	·94	·94	39
40	1·19	1·17	1·15	1·13	1·11	1·10	1·08	1·06	1·05	1·04	1·02	1·01	1·00	·99	·98	·97	40
41	1·23	1·21	1·19	1·17	1·15	1·13	1·12	1·10	1·09	1·07	1·06	1·05	1·04	1·03	1·01	1·00	41
42	1·28	1·25	1·23	1·21	1·19	1·18	1·16	1·14	1·13	1·11	1·10	1·09	1·07	1·06	1·05	1·04	42
43	1·32	1·30	1·28	1·25	1·24	1·22	1·20	1·18	1·17	1·15	1·14	1·12	1·11	1·10	1·09	1·08	43
44	1·37	1·34	1·32	1·30	1·28	1·26	1·24	1·23	1·21	1·19	1·18	1·16	1·15	1·14	1·13	1·12	44
45	1·41	1·39	1·37	1·35	1·33	1·31	1·29	1·27	1·25	1·24	1·22	1·21	1·19	1·18	1·17	1·15	45
46	1·47	1·44	1·42	1·39	1·37	1·35	1·33	1·31	1·30	1·28	1·26	1·25	1·23	1·22	1·21	1·20	46
47	1·52	1·49	1·47	1·44	1·42	1·40	1·38	1·36	1·34	1·33	1·31	1·29	1·28	1·26	1·25	1·24	47
48	1·57	1·54	1·52	1·49	1·47	1·45	1·43	1·41	1·39	1·37	1·36	1·34	1·32	1·31	1·30	1·28	48
49	1·63	1·60	1·57	1·55	1·52	1·50	1·48	1·46	1·44	1·42	1·40	1·39	1·37	1·36	1·34	1·33	49
50	1·69	1·66	1·63	1·60	1·58	1·56	1·53	1·51	1·49	1·47	1·45	1·44	1·42	1·41	1·39	1·38	50
51	1·75	1·72	1·69	1·66	1·64	1·61	1·59	1·57	1·55	1·53	1·51	1·49	1·47	1·46	1·44	1·43	51
52	1·81	1·78	1·75	1·72	1·70	1·67	1·65	1·62	1·60	1·58	1·56	1·54	1·53	1·51	1·49	1·48	52
53	1·88	1·84	1·81	1·79	1·76	1·73	1·71	1·68	1·66	1·64	1·62	1·60	1·58	1·56	1·55	1·53	53
54	1·95	1·91	1·88	1·85	1·82	1·80	1·77	1·75	1·72	1·70	1·68	1·66	1·64	1·62	1·61	1·59	54
55	2·02	1·99	1·95	1·92	1·89	1·86	1·84	1·81	1·79	1·77	1·74	1·72	1·70	1·68	1·67	1·65	55
56	2·10	2·06	2·03	2·00	1·96	1·94	1·91	1·88	1·86	1·83	1·81	1·79	1·77	1·75	1·73	1·71	56
57	2·18	2·14	2·11	2·07	2·04	2·01	1·98	1·95	1·93	1·90	1·88	1·86	1·84	1·82	1·80	1·78	57
58	2·27	2·22	2·19	2·15	2·12	2·09	2·06	2·03	2·00	1·98	1·95	1·93	1·91	1·89	1·87	1·85	58
59	2·36	2·31	2·28	2·24	2·21	2·17	2·14	2·11	2·08	2·06	2·03	2·01	1·98	1·96	1·94	1·92	59
60	2·45	2·41	2·37	2·33	2·29	2·26	2·23	2·20	2·17	2·14	2·11	2·09	2·07	2·04	2·02	2·00	60
Dec.	135° 225°	134° 226°	133° 227°	132° 228°	131° 229°	130° 230°	129° 231°	128° 232°	127° 233°	126° 234°	125° 235°	124° 236°	123° 237°	122° 238°	121° 239°	120° 240°	Dec.

B | HOUR ANGLE. | B

B—Always named the **same** as Declination.

167

A. & B. CORRECTION.

AZIMUTHS.

Lat.	·30′	·31′	·32′	·33′	·34′	·35′	·36′	·37′	·38′	·39′	·40′	·41′	·42′	·43′	·44′	·45′	Lat.
0	73·3	72·8	72·3	71·7	71·2	70·7	70·2	69·7	69·2	68·7	68·2	67·7	67·2	66·7	66·3	65·8	0
5	73·4	72·8	72·3	71·8	71·3	70·8	70·3	69·8	69·3	68·8	68·3	67·8	67·3	66·8	66·4	65·9	5
10	73·5	73·0	72·5	72·0	71·5	71·0	70·5	70·0	69·5	69·0	68·5	68·0	67·5	67·0	66·6	66·1	10
14	73·7	73·2	72·7	72·2	71·7	71·2	70·7	70·2	69·7	69·3	68·8	68·3	67·8	67·4	66·9	66·4	14
18	74·1	73·6	73·1	72·6	72·1	71·6	71·1	70·6	70·1	69·6	69·2	68·7	68·2	67·8	67·3	66·8	18
20	74·3	73·8	73·3	72·8	72·3	71·8	71·3	70·8	70·3	69·9	69·4	68·9	68·5	68·0	67·5	67·1	20
22	74·5	74·0	73·5	73·0	72·5	72·0	71·5	71·1	70·6	70·1	69·7	69·2	68·7	68·3	67·8	67·4	22
24	74·7	74·2	73·7	73·2	72·7	72·3	71·8	71·3	70·9	70·4	69·9	69·5	69·0	68·6	68·1	67·7	24
26	74·9	74·4	74·0	73·5	73·0	72·5	72·1	71·6	71·1	70·7	70·2	69·8	69·3	68·9	68·4	68·0	26
28	75·2	74·7	74·2	73·8	73·3	72·8	72·4	71·9	71·5	71·0	70·5	70·1	69·7	69·2	68·8	68·3	28
30	75·4	75·0	74·5	74·1	73·6	73·1	72·7	72·2	71·8	71·3	70·9	70·5	70·0	69·6	69·1	68·7	30
31	75·6	75·1	74·7	74·2	73·8	73·3	72·9	72·4	72·0	71·5	71·1	70·6	70·2	69·8	69·3	68·9	31
32	75·7	75·3	74·8	74·4	73·9	73·5	73·0	72·6	72·1	71·7	71·3	70·8	70·4	70·0	69·5	69·1	32
33	75·9	75·4	75·0	74·5	74·1	73·6	73·2	72·8	72·3	71·9	71·5	71·0	70·6	70·2	69·7	69·3	33
34	76·0	75·6	75·1	74·7	74·3	73·8	73·4	72·9	72·5	72·1	71·7	71·2	70·8	70·4	70·0	69·5	34
35	76·2	75·8	75·3	74·9	74·4	74·0	73·6	73·1	72·7	72·3	71·9	71·4	71·0	70·6	70·2	69·8	35
36	76·4	75·9	75·5	75·1	74·6	74·2	73·8	73·3	72·9	72·5	72·1	71·6	71·2	70·8	70·4	70·0	36
37	76·5	76·1	75·7	75·2	74·8	74·4	74·0	73·5	73·1	72·7	72·3	71·9	71·5	71·0	70·6	70·2	37
38	76·7	76·3	75·8	75·4	75·0	74·6	74·2	73·7	73·3	72·9	72·5	72·1	71·7	71·3	70·9	70·5	38
39	76·9	76·5	76·0	75·6	75·2	74·8	74·4	74·0	73·5	73·1	72·7	72·3	71·9	71·5	71·1	70·7	39
40	77·1	76·6	76·2	75·8	75·4	75·0	74·6	74·2	73·8	73·4	73·0	72·6	72·2	71·8	71·4	71·0	40
41	77·2	76·8	76·4	76·0	75·6	75·2	74·8	74·4	74·0	73·6	73·2	72·8	72·4	72·0	71·6	71·2	41
42	77·4	77·0	76·6	76·2	75·8	75·4	75·0	74·6	74·2	73·8	73·4	73·1	72·7	72·3	71·9	71·5	42
43	77·6	77·2	76·8	76·4	76·0	75·6	75·2	74·9	74·5	74·1	73·7	73·3	72·9	72·5	72·2	71·8	43
44	77·8	77·4	77·0	76·6	76·3	75·9	75·5	75·1	74·7	74·3	73·9	73·6	73·2	72·8	72·4	72·1	44
45	78·0	77·6	77·3	76·9	76·5	76·1	75·7	75·3	75·0	74·6	74·2	73·8	73·5	73·1	72·7	72·3	45
46	78·2	77·8	77·5	77·1	76·7	76·3	76·0	75·6	75·2	74·8	74·5	74·1	73·7	73·4	73·0	72·6	46
47	78·4	78·1	77·7	77·3	76·9	76·6	76·2	75·8	75·5	75·1	74·7	74·4	74·0	73·7	73·3	72·9	47
48	78·6	78·3	77·9	77·5	77·2	76·8	76·5	76·1	75·7	75·4	75·0	74·7	74·3	73·9	73·6	73·2	48
49	78·9	78·5	78·1	77·8	77·4	77·1	76·7	76·4	76·0	75·6	75·3	74·9	74·6	74·2	73·9	73·6	49
50	79·1	78·7	78·4	78·0	77·7	77·3	77·0	76·6	76·3	75·9	75·6	75·2	74·9	74·5	74·2	73·9	50
51	79·3	79·0	78·6	78·3	77·9	77·6	77·2	76·9	76·6	76·2	75·9	75·5	75·2	74·9	74·5	74·2	51
52	79·5	79·2	78·9	78·5	78·2	77·8	77·5	77·2	76·8	76·5	76·2	75·8	75·5	75·2	74·8	74·5	52
53	79·8	79·4	79·1	78·8	78·4	78·1	77·8	77·4	77·1	76·8	76·5	76·1	75·8	75·5	75·2	74·8	53
54	80·0	79·7	79·3	79·0	78·7	78·4	78·1	77·7	77·4	77·1	76·8	76·5	76·1	75·8	75·5	75·2	54
55	80·2	79·9	79·6	79·3	79·0	78·6	78·3	78·0	77·7	77·4	77·1	76·8	76·5	76·1	75·8	75·5	55
56	80·5	80·2	79·9	79·5	79·2	78·9	78·6	78·3	78·0	77·7	77·4	77·1	76·8	76·5	76·2	75·9	56
57	80·7	80·4	80·1	79·8	79·5	79·2	78·9	78·6	78·3	78·0	77·7	77·4	77·1	76·8	76·5	76·2	57
58	81·0	80·7	80·4	80·1	79·8	79·5	79·2	78·9	78·6	78·3	78·0	77·7	77·5	77·2	76·9	76·6	58
59	81·2	80·9	80·6	80·4	80·1	79·8	79·5	79·2	78·9	78·6	78·4	78·1	77·8	77·5	77·2	77·0	59
60	81·5	81·2	80·9	80·6	80·4	80·1	79·8	79·5	79·2	79·0	78·7	78·4	78·1	77·9	77·6	77·3	60
61	81·7	81·5	81·2	80·9	80·6	80·4	80·1	79·8	79·6	79·3	79·0	78·8	78·5	78·2	78·0	77·7	61
62	82·0	81·7	81·5	81·2	80·9	80·7	80·4	80·1	79·9	79·6	79·4	79·1	78·8	78·6	78·3	78·1	62
63	82·2	82·0	81·7	81·5	81·2	81·0	80·7	80·5	80·2	80·0	79·7	79·5	79·2	79·0	78·7	78·5	63
64	82·5	82·3	82·0	81·8	81·5	81·3	81·0	80·8	80·5	80·3	80·1	79·8	79·6	79·3	79·1	78·8	64
65	82·8	82·5	82·3	82·1	81·8	81·6	81·3	81·1	80·9	80·6	80·4	80·2	79·9	79·7	79·5	79·2	65
66	83·0	82·8	82·6	82·4	82·1	81·9	81·7	81·4	81·2	81·0	80·8	80·5	80·3	80·1	79·9	79·6	66
67	83·3	83·1	82·9	82·7	82·4	82·2	82·0	81·8	81·6	81·3	81·1	80·9	80·7	80·5	80·2	80·0	67
68	83·6	83·4	83·2	83·0	82·7	82·5	82·3	82·1	81·9	81·7	81·5	81·3	81·1	80·8	80·6	80·4	68

AZIMUTHS.

| Lat. | ·30′ | ·31′ | ·32′ | ·33′ | ·34′ | ·35′ | ·36′ | ·37′ | ·38′ | ·39′ | ·40′ | ·41′ | ·42′ | ·43′ | ·44′ | ·45′=A±B | Lat. |

A & B { Same names } RULE TO FIND { A & B { Different names,
    take { Sum, (add.) } C CORRECTION { take { Difference, (sub.)

C CORRECTION, (A ± B) is named the same as the greater of these quantities.

AZIMUTH takes combined names of **C** Correction and Hour Angle.

C   C

A & B CORRECTION.

AZIMUTHS.

A ± B =	·60′	·62′	·64′	·66′	·68′	·70′	·72′	·74′	·76′	·78′	·80′	·82′	·84′	·86′	·88′	·90′ = A ± B
Lat.																Lat.
0	59·0	58·2	57·4	56·6	55·8	55·0	54·2	53·5	52·8	52·0	51·3	50·6	50·0	49·3	48·7	48·0	0
5	59·1	58·3	57·5	56·7	55·9	55·1	54·3	53·6	52·9	52·2	51·5	50·8	50·1	49·4	48·8	48·1	5
10	59·4	58·6	57·8	57·0	56·2	55·4	54·7	53·9	53·2	52·5	51·8	51·1	50·4	49·7	49·1	48·4	10
14	59·8	59·0	58·2	57·4	56·6	55·8	55·1	54·3	53·6	52·9	52·2	51·5	50·8	50·2	49·5	48·9	14
18	60·3	59·5	58·7	57·9	57·1	56·3	55·6	54·9	54·1	53·4	52·7	52·1	51·4	50·7	50·1	49·4	18
20	60·6	59·8	59·0	58·2	57·4	56·7	55·9	55·2	54·5	53·8	53·1	52·4	51·7	51·1	50·4	49·8	20
22	60·9	60·1	59·3	58·5	57·8	57·0	56·3	55·5	54·8	54·1	53·4	52·8	52·1	51·4	50·8	50·2	22
24	61·3	60·5	59·7	58·9	58·2	57·4	56·7	55·9	55·2	54·5	53·8	53·2	52·5	51·8	51·2	50·6	24
26	61·7	60·9	60·1	59·3	58·6	57·8	57·1	56·4	55·7	55·0	54·3	53·6	52·9	52·3	51·7	51·0	26
28	62·1	61·3	60·5	59·8	59·0	58·3	57·6	56·8	56·1	55·4	54·8	54·1	53·4	52·8	52·2	51·5	28
30	62·5	61·8	61·0	60·2	59·5	58·8	58·1	57·3	56·6	56·0	55·3	54·6	54·0	53·3	52·7	52·1	30
31	62·8	62·0	61·3	60·5	59·8	59·0	58·3	57·6	56·9	56·2	55·6	54·9	54·2	53·6	53·0	52·4	31
32	63·0	62·3	61·5	60·8	60·0	59·3	58·6	57·9	57·2	56·5	55·8	55·2	54·5	53·9	53·3	52·7	32
33	63·3	62·5	61·8	61·0	60·3	59·6	58·9	58·2	57·5	56·8	56·1	55·5	54·8	54·2	53·6	53·0	33
34	63·6	62·8	62·1	61·3	60·6	59·9	59·2	58·5	57·8	57·1	56·4	55·8	55·1	54·5	53·9	53·3	34
35	63·8	63·1	62·3	61·6	60·9	60·2	59·5	58·8	58·1	57·4	56·8	56·1	55·5	54·8	54·2	53·6	35
36	64·1	63·4	62·6	61·9	61·2	60·5	59·8	59·1	58·4	57·7	57·1	56·4	55·8	55·2	54·6	53·9	36
37	64·4	63·7	62·9	62·2	61·5	60·8	60·1	59·4	58·7	58·1	57·4	56·8	56·1	55·5	54·9	54·3	37
38	64·7	64·0	63·2	62·5	61·8	61·1	60·4	59·8	59·1	58·4	57·8	57·1	56·5	55·9	55·3	54·7	38
39	65·0	64·3	63·6	62·8	62·1	61·5	60·8	60·1	59·4	58·8	58·1	57·5	56·9	56·2	55·6	55·0	39
40	65·3	64·6	63·9	63·2	62·5	61·8	61·1	60·5	59·8	59·1	58·5	57·9	57·2	56·6	56·0	55·4	40
41	65·6	64·9	64·2	63·5	62·8	62·2	61·5	60·8	60·2	59·5	58·9	58·2	57·6	57·0	56·4	55·8	41
42	66·0	65·3	64·6	63·9	63·2	62·5	61·9	61·2	60·5	59·9	59·3	58·6	58·0	57·4	56·8	56·2	42
43	66·3	65·6	64·9	64·2	63·6	62·9	62·2	61·6	60·9	60·3	59·7	59·0	58·4	57·8	57·2	56·6	43
44	66·7	66·0	65·3	64·6	63·9	63·3	62·6	62·0	61·3	60·7	60·1	59·5	58·9	58·3	57·7	57·1	44
45	67·0	66·3	65·7	65·0	64·3	63·7	63·0	62·4	61·7	61·1	60·5	59·9	59·3	58·7	58·1	57·6	45
46	67·4	66·7	66·0	65·4	64·7	64·1	63·4	62·8	62·2	61·5	60·9	60·3	59·7	59·1	58·6	58·0	46
47	67·7	67·1	66·4	65·8	65·1	64·5	63·8	63·2	62·6	62·0	61·4	60·8	60·2	59·6	59·0	58·5	47
48	68·1	67·5	66·8	66·2	65·5	64·9	64·3	63·7	63·0	62·4	61·8	61·2	60·7	60·1	59·5	58·9	48
49	68·5	67·9	67·2	66·6	66·0	65·3	64·7	64·1	63·5	62·9	62·3	61·7	61·1	60·6	60·0	59·4	49
50	68·9	68·3	67·6	67·0	66·4	65·8	65·2	64·6	64·0	63·4	62·8	62·2	61·6	61·1	60·5	60·0	50
51	69·3	68·7	68·1	67·4	66·8	66·2	65·6	65·0	64·4	63·9	63·3	62·7	62·1	61·6	61·0	60·5	51
52	69·7	69·1	68·5	67·9	67·3	66·7	66·1	65·5	64·9	64·4	63·8	63·2	62·7	62·1	61·6	61·0	52
53	70·1	69·5	68·9	68·3	67·7	67·2	66·6	66·0	65·4	64·9	64·3	63·7	63·2	62·6	62·1	61·6	53
54	70·6	70·0	69·4	68·8	68·2	67·6	67·1	66·5	65·9	65·4	64·8	64·3	63·7	63·2	62·6	62·1	54
55	71·0	70·4	69·8	69·3	68·7	68·1	67·6	67·0	66·4	65·9	65·4	64·8	64·3	63·7	63·2	62·7	55
56	71·5	70·9	70·3	69·7	69·2	68·6	68·1	67·5	67·0	66·4	65·9	65·4	64·9	64·3	63·8	63·3	56
57	71·9	71·3	70·8	70·2	69·7	69·1	68·6	68·0	67·5	67·0	66·5	65·9	65·4	64·9	64·4	63·9	57
58	72·4	71·8	71·3	70·7	70·2	69·6	69·1	68·6	68·1	67·5	67·0	66·5	66·0	65·5	65·0	64·5	58
59	72·8	72·3	71·8	71·2	70·7	70·2	69·7	69·1	68·6	68·1	67·6	67·1	66·6	66·1	65·6	65·1	59
60	73·3	72·8	72·3	71·7	71·2	70·7	70·2	69·7	69·2	68·7	68·2	67·7	67·2	66·7	66·3	65·8	60
61	73·8	73·3	72·8	72·3	71·8	71·3	70·8	70·3	69·8	69·3	68·8	68·3	67·8	67·4	66·9	66·4	61
62	74·3	73·8	73·3	72·8	72·3	71·8	71·3	70·8	70·4	69·9	69·4	68·9	68·5	68·0	67·6	67·1	62
63	74·8	74·3	73·8	73·3	72·8	72·4	71·9	71·4	71·0	70·5	70·0	69·6	69·1	68·7	68·2	67·8	63
64	75·3	74·8	74·3	73·9	73·4	72·9	72·5	72·0	71·6	71·1	70·7	70·2	69·8	69·3	68·9	68·5	,64
65	75·8	75·3	74·9	74·4	74·0	73·5	73·1	72·6	72·2	71·8	71·3	70·9	70·5	70·0	69·6	69·2	65
66	76·3	75·8	75·4	75·0	74·5	74·1	73·7	73·2	72·8	72·4	72·0	71·6	71·1	70·7	70·3	69·9	66
67	76·8	76·4	76·0	75·5	75·1	74·7	74·3	73·9	73·5	73·1	72·6	72·2	71·8	71·4	71·0	70·6	67
68	77·3	76·9	76·5	76·1	75·7	75·3	74·9	74·5	74·1	73·7	73·3	72·9	72·5	72·1	71·8	71·4	68
Lat.																Lat.
A ± B =	·60′	·62′	·64′	·66′	·68′	·70′	·72′	·74′	·76′	·78′	·80′	·82′	·84′	·86′	·88′	·90′ = A ± B

AZIMUTHS.

A & B Same names } RULE TO FIND { A & B Different names,
take Sum, (add.) } C CORRECTION { take Difference, (sub.)

C CORRECTION, (A ± B) is named the same as the greater of these quantities.

AZIMUTH takes combined names of **C** Correction and Hour Angle.

C   C

TRUE AMPLITUDES

Lat.	Declination														
	1°	2°	3°	4°	5°	6°	7°	8°	9°	10°	11°	12°	13°	14°	15°
°	°	°	°	°	°	°	°	°	°	°	°	°	°	°	°
2	1·0	2·0	3·0	4·0	5·0	6·0	7·0	8·0	9·0	10·0	11·0	12·0	13·0	14·0	15·0
4	1·0	2·0	3·0	4·0	5·0	6·0	7·0	8·0	9·0	10·0	11·0	12·0	13·0	14·0	15·0
6	1·0	2·0	3·0	4·0	5·0	6·0	7·1	8·1	9·1	10·1	11·1	12·1	13·1	14·1	15·1
8	1·0	2·0	3·0	4·1	5·1	6·1	7·1	8·1	9·1	10·1	11·1	12·1	13·1	14·1	15·2
10	1·0	2·0	3·1	4·1	5·1	6·1	7·1	8·1	9·2	10·2	11·2	12·2	13·2	14·2	15·3
12	1·0	2·1	3·1	4·1	5·1	6·1	7·2	8·2	9·2	10·2	11·3	12·3	13·3	14·3	15·4
14	1·0	2·1	3·1	4·1	5·2	6·2	7·2	8·3	9·3	10·3	11·3	12·4	13·4	14·4	15·5
16	1·1	2·1	3·1	4·2	5·2	6·2	7·3	8·3	9·4	10·4	11·4	12·5	13·5	14·6	15·6
18	1·1	2·1	3·2	4·2	5·3	6·3	7·4	8·4	9·5	10·5	11·6	12·6	13·7	14·7	15·8
20	1·1	2·1	3·2	4·3	5·3	6·4	7·5	8·5	9·6	10·7	11·7	12·8	13·9	14·9	15·9
22	1·1	2·2	3·2	4·3	5·4	6·5	7·6	8·6	9·7	10·8	11·9	13·0	14·1	15·1	16·2
24	1·1	2·2	3·3	4·4	5·5	6·6	7·7	8·8	9·9	11·0	12·1	13·2	14·3	15·4	16·5
26	1·1	2·2	3·4	4·5	5·6	6·7	7·8	8·9	10·0	11·2	12·3	13·4	14·5	15·6	16·8
28	1·1	2·3	3·4	4·5	5·7	6·8	7·9	9·1	10·2	11·4	12·5	13·6	14·8	15·9	17·1
30	1·2	2·3	3·5	4·6	5·8	6·9	8·1	9·3	10·4	11·6	12·7	13·9	15·1	16·2	17·4
31	1·2	2·3	3·5	4·7	5·8	7·0	8·2	9·4	10·5	11·7	12·9	14·0	15·2	16·4	17·6
32	1·2	2·4	3·6	4·7	5·9	7·1	8·3	9·5	10·6	11·8	13·0	14·2	15·4	16·6	17·8
33	1·2	2·4	3·6	4·8	6·0	7·2	8·4	9·6	10·8	12·0	13·2	14·4	15·6	16·8	18·0
34	1·2	2·4	3·6	4·8	6·0	7·3	8·5	9·7	10·9	12·1	13·3	14·5	15·8	17·0	18·2
35	1·2	2·5	3·7	4·9	6·1	7·3	8·6	9·8	11·0	12·2	13·5	14·7	16·0	17·2	18·4
36	1·2	2·5	3·7	5·0	6·2	7·4	8·7	9·9	11·2	12·4	13·7	14·9	16·2	17·4	18·7
37	1·3	2·5	3·8	5·0	6·3	7·5	8·8	10·0	11·3	12·6	13·8	15·1	16·4	17·6	18·9
38	1·3	2·5	3·8	5·1	6·4	7·6	8·9	10·2	11·4	12·7	14·0	15·3	16·6	17·9	19·2
39	1·3	2·6	3·9	5·2	6·4	7·7	9·0	10·3	11·6	12·9	14·2	15·5	16·8	18·1	19·5
40	1·3	2·6	3·9	5·2	6·5	7·9	9·2	10·5	11·8	13·1	14·4	15·8	17·1	18·4	19·8
41	1·3	2·7	4·0	5·3	6·6	8·0	9·3	10·6	12·0	13·3	14·7	16·0	17·4	18·7	20·1
42	1·4	2·7	4·0	5·4	6·7	8·1	9·4	10·8	12·2	13·5	14·9	16·3	17·6	19·0	20·4
43	1·4	2·7	4·1	5·5	6·9	8·2	9·6	11·0	12·4	13·7	15·1	16·5	17·9	19·3	20·7
44	1·4	2·8	4·2	5·6	7·0	8·4	9·8	11·2	12·6	14·0	15·4	16·8	18·2	19·7	21·1
45	1·4	2·8	4·3	5·7	7·1	8·5	9·9	11·4	12·8	14·2	15·7	17·1	18·6	20·0	21·5
46	1·4	2·9	4·3	5·8	7·2	8·7	10·1	11·6	13·0	14·5	16·0	17·4	18·9	20·4	21·9
47	1·5	2·9	4·4	5·9	7·4	8·8	10·3	11·8	13·3	14·8	16·3	17·8	19·3	20·8	22·3
48	1·5	3·0	4·5	6·0	7·5	9·0	10·5	12·0	13·5	15·1	16·6	18·1	19·7	21·2	22·8
49	1·5	3·1	4·6	6·1	7·6	9·2	10·7	12·3	13·8	15·4	16·9	18·5	20·1	21·6	23·2
50	1·6	3·1	4·7	6·2	7·8	9·4	10·9	12·5	14·1	15·7	17·3	18·9	20·5	22·1	23·8
50½	1·6	3·1	4·7	6·3	7·9	9·5	11·0	12·6	14·2	15·8	17·5	19·1	20·7	22·4	24·0
51	1·6	3·1	4·8	6·4	8·0	9·6	11·2	12·8	14·4	16·0	17·7	19·3	21·0	22·6	24·3
51½	1·6	3·2	4·8	6·4	8·0	9·7	11·3	12·9	14·6	16·2	17·8	19·5	21·2	22·9	24·6
52	1·6	3·3	4·9	6·5	8·1	9·8	11·4	13·1	14·7	16·4	18·1	19·7	21·4	23·2	24·9
52½	1·6	3·3	4·9	6·6	8·2	9·9	11·5	13·2	14·9	16·6	18·3	20·0	21·7	23·4	25·2
53	1·7	3·3	5·0	6·7	8·3	10·0	11·7	13·4	15·1	16·8	18·5	20·2	22·0	23·7	25·5
53½	1·7	3·4	5·0	6·7	8·4	10·1	11·8	13·5	15·2	17·0	18·7	20·5	22·2	24·0	25·8
54	1·7	3·4	5·1	6·8	8·5	10·3	12·0	13·7	15·4	17·2	19·0	20·7	22·5	24·3	26·1
54½	1·7	3·4	5·2	6·9	8·6	10·4	12·2	13·9	15·6	17·4	19·2	21·0	22·8	24·6	26·5
55	1·8	3·5	5·2	7·0	8·7	10·5	12·3	14·1	15·8	17·6	19·4	21·3	23·1	25·0	26·8
55½	1·8	3·5	5·3	7·1	8·9	10·6	12·4	14·2	16·0	17·9	19·7	21·5	23·4	25·3	27·2
56	1·8	3·6	5·4	7·2	9·0	10·8	12·6	14·4	16·3	18·1	20·0	21·8	23·7	25·6	27·6
56½	1·8	3·6	5·4	7·3	9·1	10·9	12·8	14·6	16·5	18·3	20·2	22·1	24·1	26·0	28·0
57	1·8	3·7	5·5	7·4	9·2	11·1	12·9	14·8	16·7	18·6	20·5	22·4	24·4	26·4	28·4
57½	1·9	3·7	5·6	7·5	9·3	11·2	13·1	15·0	16·9	18·9	20·8	22·8	24·8	26·8	28·9
58	1·9	3·8	5·7	7·6	9·5	11·4	13·3	15·2	17·2	19·1	21·1	23·1	25·1	27·2	29·2
58½	1·9	3·8	5·7	7·7	9·6	11·5	13·5	15·4	17·4	19·4	21·4	23·4	25·5	27·6	29·7
59	2·0	3·9	5·8	7·8	9·8	11·7	13·7	15·7	17·7	19·7	21·8	23·8	25·9	28·0	30·2
59½	2·0	3·9	5·9	7·9	9·9	11·9	13·9	15·9	18·0	20·0	22·1	24·2	26·3	28·5	30·7
60	2·0	4·0	6·0	8·0	10·0	12·1	14·1	16·2	18·2	20·3	22·4	24·6	26·7	28·9	31·2
60½	2·0	4·1	6·1	8·1	10·2	12·3	14·3	16·4	18·5	20·6	22·8	25·0	27·2	29·4	31·7
61	2·1	4·1	6·2	8·3	10·4	12·5	14·6	16·7	18·8	21·0	23·2	25·4	27·7	29·9	32·3
61½	2·1	4·2	6·3	8·4	10·5	12·7	14·8	17·0	19·1	21·3	23·6	25·8	28·1	30·5	32·8
62	2·1	4·3	6·4	8·6	10·7	12·9	15·1	17·3	19·5	21·7	24·0	26·3	28·6	31·0	33·5
62½	2·2	4·3	6·5	8·7	10·9	13·1	15·3	17·5	19·8	22·1	24·4	26·8	29·2	31·6	34·1

TRUE AMPLITUDES

Lat.	16°	17°	18°	19°	20°	20½°	21°	21½°	22°	22½°	23°	23½°	24°	24½°	25°
°	°	°	°	°	°	°	°	°	°	°	°	°	°	°	°
2	16·0	17·0	18·0	19·0	20·0	20·5	21·0	21·5	22·0	22·5	23·0	23·5	24·0	24·5	25·0
4	16·0	17·1	18·1	19·1	20·1	20·6	21·0	21·6	22·1	22·6	23·1	23·6	24·1	24·6	25·1
6	16·1	17·1	18·1	19·1	20·1	20·6	21·1	21·6	22·1	22·6	23·1	23·6	24·1	24·6	25·1
8	16·2	17·2	18·2	19·2	20·2	20·7	21·2	21·7	22·2	22·7	23·2	23·7	24·3	24·8	25·3
10	16·3	17·3	18·3	19·3	20·3	20·8	21·4	21·8	22·4	22·9	23·4	23·9	24·4	24·9	25·4
12	16·4	17·4	18·4	19·4	20·5	21·0	21·5	22·0	22·5	23·0	23·6	24·1	24·6	25·1	25·6
14	16·5	17·5	18·6	19·6	20·6	21·2	21·7	22·2	22·7	23·2	23·8	24·3	24·8	25·3	25·8
16	16·7	17·7	18·8	19·8	20·9	21·4	21·9	22·4	22·9	23·5	24·0	24·5	25·0	25·6	26·1
18	16·9	17·9	19·0	20·0	21·1	21·6	22·1	22·7	23·2	23·7	24·3	24·8	25·3	25·9	26·4
20	17·1	18·1	19·2	20·3	21·4	21·9	22·4	23·0	23·5	24·0	24·6	25·1	25·7	26·2	26·7
22	17·3	18·4	19·5	20·6	21·7	22·2	22·7	23·3	23·8	24·4	24·9	25·5	26·0	26·6	27·1
24	17·6	18·7	19·8	20·9	22·0	22·5	23·1	23·7	24·2	24·8	25·3	25·9	26·4	27·0	27·6
26	17·9	19·0	20·1	21·2	22·4	22·9	23·5	24·1	24·6	25·2	25·8	26·3	26·9	27·5	28·0
28	18·2	19·3	20·5	21·6	22·8	23·4	24·0	24·5	25·1	25·7	26·3	26·8	27·4	28·0	28·6
30	18·6	19·7	20·9	22·1	23·3	23·9	24·5	25·0	25·6	26·2	26·8	27·4	28·0	28·6	29·2
31	18·8	20·0	21·1	22·3	23·5	24·1	24·7	25·3	25·9	26·5	27·1	27·7	28·3	28·9	29·5
32	19·0	20·2	21·4	22·6	23·8	24·4	25·0	25·6	26·2	26·8	27·4	28·0	28·7	29·3	29·9
33	19·2	20·4	21·6	22·9	24·1	24·7	25·3	25·9	26·5	27·1	27·8	28·4	29·0	29·6	30·3
34	19·4	20·6	21·9	23·1	24·4	25·0	25·6	26·2	26·9	27·5	28·1	28·7	29·4	29·9	30·7
35	19·7	20·9	22·2	23·4	24·7	25·3	26·0	26·6	27·2	27·9	28·5	29·1	29·8	30·4	31·1
36	19·9	21·2	22·5	23·7	25·0	25·7	26·3	26·9	27·6	28·2	28·9	29·5	30·2	30·8	31·5
37	20·2	21·5	22·8	24·1	25·4	26·0	26·7	27·3	28·0	28·6	29·3	30·0	30·6	31·3	31·9
38	20·5	21·8	23·1	24·4	25·7	26·4	27·1	27·7	28·4	29·1	29·7	30·4	31·1	31·8	32·4
39	20·8	22·1	23·4	24·8	26·1	26·8	27·5	28·1	28·8	29·5	30·2	30·9	31·6	32·3	32·9
40	21·1	22·4	23·8	25·2	26·5	27·2	27·9	28·6	29·3	30·0	30·7	31·4	32·1	32·8	33·5
41	21·4	22·8	24·2	25·6	27·0	27·6	28·4	29·1	29·8	30·5	31·2	31·9	32·6	33·3	34·1
42	21·8	23·2	24·6	26·0	27·4	28·1	28·8	29·6	30·3	31·0	31·7	32·5	33·2	33·9	34·7
43	22·1	23·6	25·0	26·4	27·9	28·6	29·3	30·1	30·8	31·6	32·3	33·0	33·8	34·5	35·3
44	22·5	24·0	25·4	26·9	28·4	29·1	29·9	30·6	31·4	32·1	32·9	33·7	34·4	35·2	36·0
45	23·0	24·4	25·9	27·4	28·9	29·7	30·5	31·3	32·0	32·8	33·6	34·3	35·1	35·9	36·7
46	23·4	24·9	26·4	28·0	29·5	30·3	31·1	31·8	32·6	33·4	34·2	35·0	35·8	36·7	37·5
47	23·8	25·4	27·0	28·5	30·1	30·9	31·7	32·5	33·3	34·1	35·0	35·8	36·6	37·5	38·3
48	24·3	25·9	27·5	29·1	30·7	31·6	32·4	33·2	34·1	34·9	35·7	36·6	37·4	38·3	39·2
49	24·9	26·5	28·1	29·8	31·4	32·3	33·1	34·0	34·8	35·7	36·6	37·4	38·3	39·2	40·1
50	25·4	27·1	28·7	30·4	32·2	33·0	33·9	34·8	35·6	36·5	37·4	38·3	39·3	40·2	41·1
50½	25·7	27·4	29·1	30·8	32·5	33·4	34·3	35·2	36·1	37·0	37·9	38·8	39·8	40·7	41·6
51	26·0	27·7	29·4	31·2	32·9	33·8	34·7	35·6	36·5	37·5	38·4	39·3	40·3	41·2	42·2
51½	26·3	28·0	29·8	31·5	33·3	34·2	35·1	36·1	37·0	37·9	38·9	39·8	40·8	41·8	42·8
52	26·6	28·4	30·1	31·9	33·8	34·7	35·6	36·5	37·5	38·4	39·4	40·4	41·4	42·3	43·4
52½	26·9	28·7	30·5	32·3	34·2	35·1	36·1	37·0	38·0	38·9	39·9	40·9	41·9	42·9	44·0
53	27·3	29·1	30·9	32·8	34·6	35·6	36·6	37·5	38·5	39·5	40·5	41·5	42·5	43·6	44·6
53½	27·6	29·4	31·3	33·2	35·1	36·1	37·0	38·0	39·0	40·0	41·1	42·1	43·1	44·2	45·3
54	28·0	29·8	31·7	33·6	35·6	36·6	37·6	38·6	39·6	40·6	41·7	42·7	43·8	44·9	46·0
54½	28·3	30·2	32·2	34·1	36·1	37·1	38·1	39·1	40·1	41·2	42·3	43·4	44·5	45·6	46·7
55	28·7	30·7	32·6	34·6	36·6	37·6	38·7	39·7	40·8	41·9	42·9	44·0	45·2	46·3	47·5
55½	29·1	31·1	33·1	35·1	37·1	38·2	39·3	40·3	41·4	42·5	43·6	44·7	45·9	46·9	48·3
56	29·5	31·5	33·6	35·6	37·7	38·8	39·9	41·0	42·1	43·2	44·3	45·5	46·7	47·9	49·1
56½	30·0	32·0	34·0	36·1	38·3	39·4	40·5	41·6	42·6	43·9	45·1	46·3	47·5	48·7	50·0
57	30·4	32·5	34·6	36·7	38·9	40·0	41·2	42·3	43·5	44·6	45·9	47·1	48·3	49·6	50·9
57½	30·9	33·0	35·1	37·3	39·5	40·7	41·8	43·0	44·2	45·4	46·7	47·9	49·2	50·5	51·9
58	31·3	33·5	35·7	37·9	40·2	41·4	42·6	43·8	45·0	46·2	47·5	48·8	50·1	51·5	52·9
58½	31·8	34·0	36·3	38·5	40·9	42·1	43·3	44·5	45·8	47·1	48·4	49·7	51·1	52·5	54·0
59	32·4	34·6	36·9	39·2	41·6	42·8	44·1	45·4	46·7	48·0	49·4	50·7	52·2	53·6	55·1
59½	32·9	35·2	37·5	39·9	42·4	43·6	44·9	46·2	47·6	48·9	50·3	51·8	53·3	54·8	56·4
60	33·5	35·8	38·2	40·6	43·2	44·5	45·8	47·1	48·5	49·9	51·4	52·9	54·4	56·0	57·7
60½	34·0	36·4	38·8	41·4	44·0	45·3	46·7	48·1	49·5	51·0	52·5	54·1	55·7	57·4	59·1
61	34·7	37·1	39·6	42·2	44·9	46·3	47·7	49·1	50·6	52·1	53·7	55·3	57·0	58·8	60·6
61½	35·3	37·8	40·4	43·0	45·8	47·2	48·7	50·2	51·7	53·3	55·0	56·7	58·5	60·4	62·3
62	36·0	38·5	41·2	43·9	46·8	48·2	49·8	51·3	52·9	54·6	56·3	58·1	60·0	62·0	64·2
62½	36·7	39·3	42·0	44·8	47·8	49·3	50·9	52·5	54·2	56·0	57·8	59·7	61·7	63·9	66·2

A2 ALTITUDE CORRECTION TABLES 10°–90°—SUN, STARS, PLANETS

SUN

App. Alt.	OCT.—MAR. Lower Limb	Upper Limb	App. Alt.	APR.—SEPT. Lower Limb	Upper Limb
9 34	+10.8	−21.5	9 39	+10.6	21.2
9 45	+10.9	21.4	9 51	+10.7	21.1
9 56	+11.0	21.3	10 03	+10.8	21.0
10 08	+11.1	21.2	10 15	+10.9	20.9
10 21	+11.2	21.1	10 27	+11.0	20.8
10 34	+11.3	21.0	10 40	+11.1	20.7
10 47	+11.4	20.9	10 54	+11.2	20.6
11 01	+11.5	20.8	11 08	+11.3	20.5
11 15	+11.6	20.7	11 23	+11.4	20.4
11 30	+11.7	20.6	11 38	+11.5	20.3
11 46	+11.8	20.5	11 54	+11.6	20.2
12 02	+11.9	20.4	12 10	+11.7	20.1
12 19	+12.0	20.3	12 28	+11.8	20.0
12 37	+12.1	20.2	12 46	+11.9	19.9
12 55	+12.2	20.1	13 05	+12.0	19.8
13 14	+12.3	20.0	13 24	+12.1	19.7
13 35	+12.4	19.9	13 45	+12.2	19.6
13 56	+12.5	19.8	14 07	+12.3	19.5
14 18	+12.6	19.7	14 30	+12.4	19.4
14 42	+12.7	19.6	14 54	+12.5	19.3
15 06	+12.8	19.5	15 19	+12.6	19.2
15 32	+12.9	19.4	15 46	+12.7	19.1
15 59	+13.0	19.3	16 14	+12.8	19.0
16 28	+13.1	19.2	16 44	+12.9	18.9
16 59	+13.2	19.1	17 15	+13.0	18.8
17 32	+13.3	19.0	17 48	+13.1	18.7
18 06	+13.4	18.9	18 24	+13.2	18.6
18 42	+13.5	18.8	19 01	+13.3	18.5
19 21	+13.6	18.7	19 42	+13.4	18.4
20 03	+13.7	18.6	20 25	+13.5	18.3
20 48	+13.8	18.5	21 11	+13.6	18.2
21 35	+13.9	18.4	22 00	+13.7	18.1
22 26	+14.0	18.3	22 54	+13.8	18.0
23 22	+14.1	18.2	23 51	+13.9	17.9
24 21	+14.2	18.1	24 53	+14.0	17.8
25 26	+14.3	18.0	26 00	+14.1	17.7
26 36	+14.4	17.9	27 13	+14.2	17.6
27 52	+14.5	17.8	28 33	+14.3	17.5
29 15	+14.6	17.7	30 00	+14.4	17.4
30 46	+14.7	17.6	31 35	+14.5	17.3
32 26	+14.8	17.5	33 20	+14.6	17.2
34 17	+14.9	17.4	35 17	+14.7	17.1
36 20	+15.0	17.3	37 26	+14.8	17.0
38 36	+15.1	17.2	39 50	+14.9	16.9
41 08	+15.2	17.1	42 31	+15.0	16.8
43 59	+15.3	17.0	45 31	+15.1	16.7
47 10	+15.4	16.9	48 55	+15.2	16.6
50 46	+15.5	16.8	52 44	+15.3	16.5
54 49	+15.6	16.7	57 02	+15.4	16.4
59 23	+15.7	16.6	61 51	+15.5	16.3
64 30	+15.8	16.5	67 17	+15.6	16.2
70 12	+15.9	16.4	73 16	+15.7	16.1
76 26	+16.0	16.3	79 43	+15.8	16.0
83 05	+16.1	16.2	86 32	+15.9	15.9
90 00			90 00		

STARS AND PLANETS

App. Alt.	Corrn
9 56	−5.3
10 08	−5.2
10 20	−5.1
10 33	−5.0
10 46	−4.9
11 00	−4.8
11 14	−4.7
11 29	−4.6
11 45	−4.5
12 01	−4.4
12 18	−4.3
12 35	−4.2
12 54	−4.1
13 13	−4.0
13 33	−3.9
13 54	−3.8
14 16	−3.7
14 40	−3.6
15 04	−3.5
15 30	−3.4
15 57	−3.3
16 26	−3.2
16 56	−3.1
17 28	−3.0
18 02	−2.9
18 38	−2.8
19 17	−2.7
19 58	−2.6
20 42	−2.5
21 28	−2.4
22 19	−2.3
23 13	−2.2
24 11	−2.1
25 14	−2.0
26 22	−1.9
27 36	−1.8
28 56	−1.7
30 24	−1.6
32 00	−1.5
33 45	−1.4
35 40	−1.3
37 48	−1.2
40 08	−1.1
42 44	−1.0
45 36	−0.9
48 47	−0.8
52 18	−0.7
56 11	−0.6
60 28	−0.5
65 08	−0.4
70 11	−0.3
75 34	−0.2
81 13	−0.1
87 03	0.0
90 00	0.0

Additional Corrn — 1972

App. Alt.	Corrn
VENUS	
Jan.1–Feb.29	
0°	
42	+0.1
Mar.1–Apr.15	
0°	
47	+0.2
Apr.16–May12	
0°	
46	+0.3
May13–May27	
0°	
11	+0.4
41	+0.5
May28–June5	
0°	
6	+0.5
20	+0.6
31	+0.7
June6–June29	
0°	
4	+0.6
12	+0.7
22	+0.8
June30–July8	
0°	
6	+0.5
20	+0.6
31	+0.7
July9–July24	
0°	
11	+0.4
41	+0.5
July25–Aug.19	
0°	
46	+0.3
Aug.20–Oct.5	
0°	
47	+0.2
Oct.6–Dec.31	
0°	
42	+0.1
MARS	
Jan.1–Dec.31	
0°	
60	+0.1

DIP

Ht. of Eye (m)	Corrn	Ht. of Eye (ft)	Ht. of Eye (m)	Corrn
2.4	−2.8	8.0	1.0	−1.8
2.6	−2.9	8.6	1.5	−2.2
2.8	−3.0	9.2	2.0	−2.5
3.0	−3.1	9.8	2.5	−2.8
3.2	−3.2	10.5	3.0	−3.0
3.4	−3.3	11.2	See table	
3.6	−3.4	11.9	←	
3.8	−3.5	12.6		
4.0	−3.6	13.3	m	
4.3	−3.7	14.1	20	−7.9
4.5	−3.8	14.9	22	−8.3
4.7	−3.9	15.7	24	−8.6
5.0	−4.0	16.5	26	−9.0
5.2	−4.1	17.4	28	−9.3
5.5	−4.2	18.3		
5.8	−4.3	19.1	30	−9.6
6.1	−4.4	20.1	32	−10.0
6.3	−4.5	21.0	34	−10.3
6.6	−4.6	22.0	36	−10.6
6.9	−4.7	22.9	38	−10.8
7.2	−4.8	23.9		
7.5	−4.9	24.9	40	−11.1
7.9	−5.0	26.0	42	−11.4
8.2	−5.1	27.1	44	−11.7
8.5	−5.2	28.1	46	−11.9
8.8	−5.3	29.2	48	−12.2
9.2	−5.4	30.4		
9.5	−5.5	31.5	ft.	
9.9	−5.6	32.7	2	−1.4
10.3	−5.7	33.9	4	−1.9
10.6	−5.8	35.1	6	−2.4
11.0	−5.9	36.3	8	−2.7
11.4	−6.0	37.6	10	−3.1
11.8	−6.1	38.9	See table	
12.2	−6.2	40.1	←	
12.6	−6.3	41.5	ft.	
13.0	−6.4	42.8	70	−8.1
13.4	−6.5	44.2	75	−8.4
13.8	−6.6	45.5	80	−8.7
14.2	−6.7	46.9	85	−8.9
14.7	−6.8	48.4	90	−9.2
15.1	−6.9	49.8	95	−9.5
15.5	−7.0	51.3	100	−9.7
16.0	−7.1	52.8	105	−9.9
16.5	−7.2	54.3	110	−10.2
16.9	−7.3	55.8	115	−10.4
17.4	−7.4	57.4	120	−10.6
17.9	−7.5	58.9	125	−10.8
18.4	−7.6	60.5		
18.8	−7.7	62.1	130	−11.1
19.3	−7.8	63.8	135	−11.3
19.8	−7.9	65.4	140	−11.5
20.4	−8.0	67.1	145	−11.7
20.9	−8.1	68.8	150	−11.9
21.4		70.5	155	−12.1

App. Alt. = Apparent altitude = Sextant altitude corrected for index error and dip.
For daylight observations of Venus, see page 260.

172

G.M.T.	SUN G.H.A.	SUN Dec.	MOON G.H.A.	v	MOON Dec.	d	H.P.
d h	° '	° '	° '	'	° '	'	'
3 00	180 46.9	N15 38.4	309 00.6	9.8	S25 44.3	2.9	54.9
01	195 46.9	39.1	323 29.4	9.7	25 41.4	3.0	55.0
02	210 47.0	39.9	337 58.1	9.8	25 38.4	3.1	55.0
03	225 47.1	.. 40.6	352 26.9	9.8	25 35.3	3.3	55.0
04	240 47.1	41.4	6 55.7	9.8	25 32.0	3.4	55.0
05	255 47.2	42.1	21 24.5	9.8	25 28.6	3.5	55.0
06	270 47.3	N15 42.8	35 53.3	9.8	S25 25.1	3.6	55.1
W 07	285 47.3	43.6	50 22.1	9.9	25 21.5	3.8	55.1
E 08	300 47.4	44.3	64 51.0	9.8	25 17.7	3.9	55.1
D 09	315 47.5	.. 45.0	79 19.8	9.9	25 13.8	4.0	55.1
N 10	330 47.5	45.8	93 48.7	9.8	25 09.8	4.2	55.1
E 11	345 47.6	46.5	108 17.5	9.9	25 05.6	4.3	55.2
S 12	0 47.6	N15 47.2	122 46.4	9.9	S25 01.3	4.4	55.2
D 13	15 47.7	47.9	137 15.3	10.0	24 56.9	4.6	55.2
A 14	30 47.8	48.7	151 44.3	9.9	24 52.3	4.7	55.2
Y 15	45 47.8	.. 49.4	166 13.2	10.0	24 47.6	4.8	55.2
16	60 47.9	50.1	180 42.2	9.9	24 42.8	4.9	55.3
17	75 48.0	50.9	195 11.1	10.0	24 37.9	5.1	55.3
18	90 48.0	N15 51.6	209 40.1	10.1	S24 32.8	5.2	55.3
19	105 48.1	52.3	224 09.2	10.0	24 27.6	5.3	55.3
20	120 48.1	53.0	238 38.2	10.1	24 22.3	5.4	55.4
21	135 48.2	.. 53.8	253 07.3	10.1	24 16.9	5.6	55.4
22	150 48.3	54.5	267 36.4	10.1	24 11.3	5.7	55.4
23	165 48.3	55.2	282 05.5	10.1	24 05.6	5.8	55.4
4 00	180 48.4	N15 56.0	296 34.6	10.1	S23 59.8	6.0	55.5
01	195 48.5	56.7	311 03.7	10.2	23 53.8	6.1	55.5
02	210 48.5	57.4	325 32.9	10.2	23 47.7	6.2	55.5
03	225 48.6	58.1	340 02.1	10.3	23 41.5	6.3	55.5
04	240 48.6	58.9	354 31.4	10.2	23 35.2	6.4	55.5
05	255 48.7	15 59.6	9 00.6	10.3	23 28.8	6.5	55.5
06	270 48.8	N16 00.3	23 29.9	10.3	S23 22.2	6.7	55.6
T 07	285 48.8	01.0	37 59.2	10.4	23 15.5	6.8	55.6
H 08	300 48.9	01.7	52 28.6	10.4	23 08.7	6.9	55.6
U 09	315 48.9	.. 02.5	66 58.0	10.4	23 01.8	7.1	55.7
R 10	330 49.0	03.2	81 27.4	10.4	22 54.7	7.1	55.7
S 11	345 49.0	03.9	95 56.8	10.4	22 47.6	7.3	55.7
D 12	0 49.1	N16 04.6	110 26.2	10.5	S22 40.3	7.4	55.8
A 13	15 49.2	05.3	124 55.7	10.5	22 32.9	7.6	55.8
Y 14	30 49.2	06.1	139 25.2	10.6	22 25.3	7.6	55.8
15	45 49.3	.. 06.8	153 54.8	10.6	22 17.7	7.8	55.8
16	60 49.3	07.5	168 24.4	10.6	22 09.9	7.9	55.9
17	75 49.4	08.2	182 54.0	10.6	22 02.0	8.1	55.9
18	90 49.4	N16 08.9	197 23.6	10.7	S21 54.0	8.1	55.9
19	105 49.5	09.7	211 53.3	10.7	21 45.9	8.2	55.9
20	120 49.6	10.4	226 23.0	10.7	21 37.7	8.3	56.0
21	135 49.6	.. 11.1	240 52.7	10.8	21 29.4	8.5	56.0
22	150 49.7	11.8	255 22.5	10.8	21 20.9	8.6	56.0
23	165 49.7	12.5	269 52.3	10.8	21 12.3	8.7	56.1
5 00	180 49.8	N16 13.2	284 22.1	10.9	S21 03.6	8.8	56.1
01	195 49.8	13.9	298 52.0	10.9	20 54.8	8.9	56.1
02	210 49.9	14.7	313 21.9	10.9	20 45.9	9.0	56.1
03	225 49.9	.. 15.4	327 51.8	11.0	20 36.9	9.1	56.2
04	240 50.0	16.1	342 21.8	11.0	20 27.8	9.3	56.2
05	255 50.0	16.8	356 51.8	11.0	20 18.5	9.3	56.2
06	270 50.1	N16 17.5	11 21.8	11.1	S20 09.2	9.5	56.3
07	285 50.2	18.2	25 51.9	11.0	19 59.7	9.6	56.3
08	300 50.2	19.0	40 21.9	11.2	19 50.1	9.6	56.3
F 09	315 50.3	.. 19.6	54 52.1	11.1	19 40.5	9.8	56.4
R 10	330 50.3	20.4	69 22.2	11.2	19 30.7	9.9	56.4
I 11	345 50.4	21.1	83 52.4	11.2	19 20.8	10.0	56.4
D 12	0 50.4	N16 21.8	98 22.6	11.3	S19 10.8	10.1	56.4
A 13	15 50.5	22.5	112 52.9	11.2	19 00.7	10.2	56.5
Y 14	30 50.5	23.2	127 23.1	11.3	18 50.5	10.3	56.5
15	45 50.6	.. 23.9	141 53.4	11.4	18 40.2	10.4	56.5
16	60 50.6	24.6	156 23.8	11.3	18 29.8	10.5	56.6
17	75 50.7	25.3	170 54.1	11.4	18 19.3	10.6	56.6
18	90 50.7	N16 26.0	185 24.5	11.5	S18 08.7	10.7	56.6
19	105 50.8	26.7	199 55.0	11.4	17 58.0	10.8	56.7
20	120 50.8	27.4	214 25.4	11.5	17 47.2	11.0	56.7
21	135 50.9	.. 28.1	228 55.9	11.5	17 36.2	11.0	56.7
22	150 50.9	28.8	243 26.4	11.6	17 25.2	11.1	56.8
23	165 51.0	29.5	257 57.0	11.5	17 14.1	11.2	56.8
	S.D. 15.9	d 0.7	S.D. 15.0		15.2		15.4

Lat.	Twilight Naut.	Civil	Sun-rise	Moonrise 3	4	5	6
°	h m	h m	h m	h m	h m	h m	h m
N 72	////	////	01 24	■	■	■	04 17
N 70	////	////	02 13	■	■	03 45	03 35
68	////	00 17	02 43	■	■	03 45	03 06
66	////	01 36	03 06	■	03 44	03 01	02 44
64	////	02 11	03 23	02 52	02 39	02 32	02 27
62	00 14	02 36	03 38	01 48	02 04	02 10	02 13
60	01 25	02 55	03 50	01 14	01 38	01 52	02 00
N 58	01 57	03 11	04 00	00 48	01 18	01 37	01 50
56	02 20	03 24	04 09	00 28	01 01	01 24	01 40
54	02 39	03 35	04 18	00 12	00 47	01 13	01 32
52	02 54	03 45	04 25	24 34	00 34	01 03	01 25
50	03 06	03 54	04 31	24 23	00 23	00 54	01 18
45	03 32	04 12	04 45	24 00	00 00	00 35	01 04
N 40	03 51	04 27	04 56	23 42	24 19	00 19	00 52
35	04 06	04 39	05 06	23 26	24 06	00 06	00 41
30	04 19	04 49	05 15	23 12	23 54	24 32	00 32
20	04 39	05 06	05 29	22 49	23 35	24 17	00 17
N 10	04 54	05 20	05 42	22 29	23 17	24 03	00 03
0	05 07	05 32	05 53	22 10	23 01	23 50	24 39
S 10	05 18	05 43	06 05	21 52	22 45	23 37	24 30
20	05 28	05 54	06 17	21 33	22 27	23 24	24 20
30	05 37	06 06	06 31	21 08	22 07	23 08	24 09
35	05 42	06 12	06 39	20 54	21 55	22 59	24 03
40	05 47	06 19	06 48	20 38	21 42	22 48	23 56
45	05 52	06 27	06 58	20 19	21 26	22 36	23 47
S 50	05 58	06 36	07 11	19 55	21 06	22 20	23 37
52	06 00	06 41	07 17	19 44	20 56	22 13	23 32
54	06 03	06 45	07 23	19 31	20 45	22 05	23 27
56	06 06	06 50	07 31	19 16	20 33	21 56	23 21
58	06 08	06 55	07 39	18 58	20 19	21 46	23 15
S 60	06 11	07 02	07 48	18 36	20 02	21 34	23 07

Lat.	Sun-set	Twilight Civil	Naut.	Moonset 3	4	5	6
°	h m	h m	h m	h m	h m	h m	h m
N 72	22 38	////	////	■	■	■	08 09
N 70	21 46	////	////	■	■	06 56	09 17
68	21 14	////	////	■	05 10	07 38	09 37
66	20 51	22 24	////	04 15	06 15	08 07	09 53
64	20 33	21 47	////	05 19	06 50	08 28	10 07
62	20 18	21 21	////	05 55	07 15	08 45	10 18
60	20 05	21 02	22 34	06 18	07 35	09 00	10 28
N 58	19 55	20 45	22 00	06 38	07 51	09 12	10 36
56	19 45	20 31	21 36	06 54	08 05	09 23	10 44
54	19 37	20 20	21 17	07 09	08 17	09 32	10 51
52	19 30	20 09	21 02	07 21	08 28	09 41	10 57
50	19 23	20 01	20 49	07 32	08 37	09 48	11 03
45	19 09	19 42	20 23	07 47	08 50	09 59	11 10
N 40	18 58	19 27	20 04	08 07	09 08	10 13	11 20
35	18 48	19 15	19 48	08 24	09 23	10 26	11 29
30	18 39	19 05	19 35	08 38	09 36	10 36	11 37
20	18 25	18 48	19 15	09 03	09 58	10 54	11 51
N 10	18 12	18 34	19 00	09 24	10 17	11 10	12 03
0	18 00	18 22	18 47	09 44	10 35	11 25	12 13
S 10	17 49	18 10	18 36	10 03	10 52	11 39	12 24
20	17 36	17 59	18 25	10 24	11 11	11 55	12 36
30	17 22	17 47	18 16	10 49	11 33	12 13	12 49
35	17 14	17 41	18 11	11 03	11 45	12 23	12 56
40	17 04	17 34	18 06	11 19	12 00	12 34	13 05
45	16 55	17 26	18 01	11 39	12 17	12 48	13 15
S 50	16 42	17 16	17 55	12 03	12 38	13 05	13 27
52	16 36	17 12	17 53	12 15	12 48	13 12	13 32
54	16 29	17 08	17 50	12 28	12 59	13 21	13 38
56	16 21	17 03	17 47	12 44	13 11	13 31	13 45
58	16 14	16 57	17 44	13 02	13 26	13 42	13 52
S 60	16 04	16 51	17 41	13 25	13 43	13 54	14 01

Day	SUN Eqn. of Time 00h	12h	Mer. Pass.	MOON Mer. Pass. Upper	Lower	Age	Phase
	m s	m s	h m	h m	h m	d	
3	03 07	03 10	11 57	03 31	15 57	20	
4	03 13	03 16	11 57	04 23	16 48	21	◗
5	03 19	03 22	11 57	05 13	17 38	22	

CONVERSION OF ARC TO TIME

0°–59°		60°–119°		120°–179°		180°–239°		240°–299°		300°–359°			0′·00	0′·25	0′·50	0′·75
°	h m	°	h m	°	h m	°	h m	°	h m	°	h m	′	m s	m s	m s	m s
0	0 00	60	4 00	120	8 00	180	12 00	240	16 00	300	20 00	0	0 00	0 01	0 02	0 03
1	0 04	61	4 04	121	8 04	181	12 04	241	16 04	301	20 04	1	0 04	0 05	0 06	0 07
2	0 08	62	4 08	122	8 08	182	12 08	242	16 08	302	20 08	2	0 08	0 09	0 10	0 11
3	0 12	63	4 12	123	8 12	183	12 12	243	16 12	303	20 12	3	0 12	0 13	0 14	0 15
4	0 16	64	4 16	124	8 16	184	12 16	244	16 16	304	20 16	4	0 16	0 17	0 18	0 19
5	0 20	65	4 20	125	8 20	185	12 20	245	16 20	305	20 20	5	0 20	0 21	0 22	0 23
6	0 24	66	4 24	126	8 24	186	12 24	246	16 24	306	20 24	6	0 24	0 25	0 26	0 27
7	0 28	67	4 28	127	8 28	187	12 28	247	16 28	307	20 28	7	0 28	0 29	0 30	0 31
8	0 32	68	4 32	128	8 32	188	12 32	248	16 32	308	20 32	8	0 32	0 33	0 34	0 35
9	0 36	69	4 36	129	8 36	189	12 36	249	16 36	309	20 36	9	0 36	0 37	0 38	0 39
10	0 40	70	4 40	130	8 40	190	12 40	250	16 40	310	20 40	10	0 40	0 41	0 42	0 43
11	0 44	71	4 44	131	8 44	191	12 44	251	16 44	311	20 44	11	0 44	0 45	0 46	0 47
12	0 48	72	4 48	132	8 48	192	12 48	252	16 48	312	20 48	12	0 48	0 49	0 50	0 51
13	0 52	73	4 52	133	8 52	193	12 52	253	16 52	313	20 52	13	0 52	0 53	0 54	0 55
14	0 56	74	4 56	134	8 56	194	12 56	254	16 56	314	20 56	14	0 56	0 57	0 58	0 59
15	1 00	75	5 00	135	9 00	195	13 00	255	17 00	315	21 00	15	1 00	1 01	1 02	1 03
16	1 04	76	5 04	136	9 04	196	13 04	256	17 04	316	21 04	16	1 04	1 05	1 06	1 07
17	1 08	77	5 08	137	9 08	197	13 08	257	17 08	317	21 08	17	1 08	1 09	1 10	1 11
18	1 12	78	5 12	138	9 12	198	13 12	258	17 12	318	21 12	18	1 12	1 13	1 14	1 15
19	1 16	79	5 16	139	9 16	199	13 16	259	17 16	319	21 16	19	1 16	1 17	1 18	1 19
20	1 20	80	5 20	140	9 20	200	13 20	260	17 20	320	21 20	20	1 20	1 21	1 22	1 23
21	1 24	81	5 24	141	9 24	201	13 24	261	17 24	321	21 24	21	1 24	1 25	1 26	1 27
22	1 28	82	5 28	142	9 28	202	13 28	262	17 28	322	21 28	22	1 28	1 29	1 30	1 31
23	1 32	83	5 32	143	9 32	203	13 32	263	17 32	323	21 32	23	1 32	1 33	1 34	1 35
24	1 36	84	5 36	144	9 36	204	13 36	264	17 36	324	21 36	24	1 36	1 37	1 38	1 39
25	1 40	85	5 40	145	9 40	205	13 40	265	17 40	325	21 40	25	1 40	1 41	1 42	1 43
26	1 44	86	5 44	146	9 44	206	13 44	266	17 44	326	21 44	26	1 44	1 45	1 46	1 47
27	1 48	87	5 48	147	9 48	207	13 48	267	17 48	327	21 48	27	1 48	1 49	1 50	1 51
28	1 52	88	5 52	148	9 52	208	13 52	268	17 52	328	21 52	28	1 52	1 53	1 54	1 55
29	1 56	89	5 56	149	9 56	209	13 56	269	17 56	329	21 56	29	1 56	1 57	1 58	1 59
30	2 00	90	6 00	150	10 00	210	14 00	270	18 00	330	22 00	30	2 00	2 01	2 02	2 03
31	2 04	91	6 04	151	10 04	211	14 04	271	18 04	331	22 04	31	2 04	2 05	2 06	2 07
32	2 08	92	6 08	152	10 08	212	14 08	272	18 08	332	22 08	32	2 08	2 09	2 10	2 11
33	2 12	93	6 12	153	10 12	213	14 12	273	18 12	333	22 12	33	2 12	2 13	2 14	2 15
34	2 16	94	6 16	154	10 16	214	14 16	274	18 16	334	22 16	34	2 16	2 17	2 18	2 19
35	2 20	95	6 20	155	10 20	215	14 20	275	18 20	335	22 20	35	2 20	2 21	2 22	2 23
36	2 24	96	6 24	156	10 24	216	14 24	276	18 24	336	22 24	36	2 24	2 25	2 26	2 27
37	2 28	97	6 28	157	10 28	217	14 28	277	18 28	337	22 28	37	2 28	2 29	2 30	2 31
38	2 32	98	6 32	158	10 32	218	14 32	278	18 32	338	22 32	38	2 32	2 33	2 34	2 35
39	2 36	99	6 36	159	10 36	219	14 36	279	18 36	339	22 36	39	2 36	2 37	2 38	2 39
40	2 40	100	6 40	160	10 40	220	14 40	280	18 40	340	22 40	40	2 40	2 41	2 42	2 43
41	2 44	101	6 44	161	10 44	221	14 44	281	18 44	341	22 44	41	2 44	2 45	2 46	2 47
42	2 48	102	6 48	162	10 48	222	14 48	282	18 48	342	22 48	42	2 48	2 49	2 50	2 51
43	2 52	103	6 52	163	10 52	223	14 52	283	18 52	343	22 52	43	2 52	2 53	2 54	2 55
44	2 56	104	6 56	164	10 56	224	14 56	284	18 56	344	22 56	44	2 56	2 57	2 58	2 59
45	3 00	105	7 00	165	11 00	225	15 00	285	19 00	345	23 00	45	3 00	3 01	3 02	3 03
46	3 04	106	7 04	166	11 04	226	15 04	286	19 04	346	23 04	46	3 04	3 05	3 06	3 07
47	3 08	107	7 08	167	11 08	227	15 08	287	19 08	347	23 08	47	3 08	3 09	3 10	3 11
48	3 12	108	7 12	168	11 12	228	15 12	288	19 12	348	23 12	48	3 12	3 13	3 14	3 15
49	3 16	109	7 16	169	11 16	229	15 16	289	19 16	349	23 16	49	3 16	3 17	3 18	3 19
50	3 20	110	7 20	170	11 20	230	15 20	290	19 20	350	23 20	50	3 20	3 21	3 22	3 23
51	3 24	111	7 24	171	11 24	231	15 24	291	19 24	351	23 24	51	3 24	3 25	3 26	3 27
52	3 28	112	7 28	172	11 28	232	15 28	292	19 28	352	23 28	52	3 28	3 29	3 30	3 31
53	3 32	113	7 32	173	11 32	233	15 32	293	19 32	353	23 32	53	3 32	3 33	3 34	3 35
54	3 36	114	7 36	174	11 36	234	15 36	294	19 36	354	23 36	54	3 36	3 37	3 38	3 39
55	3 40	115	7 40	175	11 40	235	15 40	295	19 40	355	23 40	55	3 40	3 41	3 42	3 43
56	3 44	116	7 44	176	11 44	236	15 44	296	19 44	356	23 44	56	3 44	3 45	3 46	3 47
57.	3 48	117	7 48	177	11 48	237	15 48	297	19 48	357	23 48	57	3 48	3 49	3 50	3 51
58	3 52	118	7 52	178	11 52	238	15 52	298	19 52	358	23 52	58	3 52	3 53	3 54	3 55
59	3 56	119	7 56	179	11 56	239	15 56	299	19 56	359	23 56	59	3 56	3 57	3 58	3 59

The above table is for converting expressions in arc to their equivalent in time ; its main use in this Almanac is for the conversion of longitude for application to L.M.T. (*added* if *west*, *subtracted* if *east*) to give G.M.T. or vice versa, particularly in the case of sunrise, sunset, etc.

44	SUN PLANETS	ARIES	MOON	v or d	Corrⁿ	v or d	Corrⁿ	v or d	Corrⁿ
s	° ′	° ′	° ′	′	′	′	′	′	′
00	11 00·0	11 01·8	10 29·9	0·0	0·0	6·0	4·5	12·0	8·9
01	11 00·3	11 02·1	10 30·2	0·1	0·1	6·1	4·5	12·1	9·0
02	11 00·5	11 02·3	10 30·4	0·2	0·1	6·2	4·6	12·2	9·0
03	11 00·8	11 02·6	10 30·6	0·3	0·2	6·3	4·7	12·3	9·1
04	11 01·0	11 02·8	10 30·9	0·4	0·3	6·4	4·7	12·4	9·2
05	11 01·3	11 03·1	10 31·1	0·5	0·4	6·5	4·8	12·5	9·3
06	11 01·5	11 03·3	10 31·4	0·6	0·4	6·6	4·9	12·6	9·3
07	11 01·8	11 03·6	10 31·6	0·7	0·5	6·7	5·0	12·7	9·4
08	11 02·0	11 03·8	10 31·8	0·8	0·6	6·8	5·0	12·8	9·5
09	11 02·3	11 04·1	10 32·1	0·9	0·7	6·9	5·1	12·9	9·6
10	11 02·5	11 04·3	10 32·3	1·0	0·7	7·0	5·2	13·0	9·6
11	11 02·8	11 04·6	10 32·6	1·1	0·8	7·1	5·3	13·1	9·7
12	11 03·0	11 04·8	10 32·8	1·2	0·9	7·2	5·3	13·2	9·8
13	11 03·3	11 05·1	10 33·0	1·3	1·0	7·3	5·4	13·3	9·9
14	11 03·5	11 05·3	10 33·3	1·4	1·0	7·4	5·5	13·4	9·9
15	11 03·8	11 05·6	10 33·5	1·5	1·1	7·5	5·6	13·5	10·0
16	11 04·0	11 05·8	10 33·8	1·6	1·2	7·6	5·6	13·6	10·1
17	11 04·3	11 06·1	10 34·0	1·7	1·3	7·7	5·7	13·7	10·2
18	11 04·5	11 06·3	10 34·2	1·8	1·3	7·8	5·8	13·8	10·2
19	11 04·8	11 06·6	10 34·5	1·9	1·4	7·9	5·9	13·9	10·3
20	11 05·0	11 06·8	10 34·7	2·0	1·5	8·0	5·9	14·0	10·4
21	11 05·3	11 07·1	10 34·9	2·1	1·6	8·1	6·0	14·1	10·5
22	11 05·5	11 07·3	10 35·2	2·2	1·6	8·2	6·1	14·2	10·5
23	11 05·8	11 07·6	10 35·4	2·3	1·7	8·3	6·2	14·3	10·6
24	11 06·0	11 07·8	10 35·7	2·4	1·8	8·4	6·2	14·4	10·7
25	11 06·3	11 08·1	10 35·9	2·5	1·9	8·5	6·3	14·5	10·8
26	11 06·5	11 08·3	10 36·1	2·6	1·9	8·6	6·4	14·6	10·8
27	11 06·8	11 08·6	10 36·4	2·7	2·0	8·7	6·5	14·7	10·9
28	11 07·0	11 08·8	10 36·6	2·8	2·1	8·8	6·5	14·8	11·0
29	11 07·3	11 09·1	10 36·9	2·9	2·2	8·9	6·6	14·9	11·1
30	11 07·5	11 09·3	10 37·1	3·0	2·2	9·0	6·7	15·0	11·1
31	11 07·8	11 09·6	10 37·3	3·1	2·3	9·1	6·7	15·1	11·2
32	11 08·0	11 09·8	10 37·6	3·2	2·4	9·2	6·8	15·2	11·3
33	11 08·3	11 10·1	10 37·8	3·3	2·4	9·3	6·9	15·3	11·3
34	11 08·5	11 10·3	10 38·0	3·4	2·5	9·4	7·0	15·4	11·4
35	11 08·8	11 10·6	10 38·3	3·5	2·6	9·5	7·0	15·5	11·5
36	11 09·0	11 10·8	10 38·5	3·6	2·7	9·6	7·1	15·6	11·6
37	11 09·3	11 11·1	10 38·8	3·7	2·7	9·7	7·2	15·7	11·6
38	11 09·5	11 11·3	10 39·0	3·8	2·8	9·8	7·3	15·8	11·7
39	11 09·8	11 11·6	10 39·2	3·9	2·9	9·9	7·3	15·9	11·8
40	11 10·0	11 11·8	10 39·5	4·0	3·0	10·0	7·4	16·0	11·9
41	11 10·3	11 12·1	10 39·7	4·1	3·0	10·1	7·5	16·1	11·9
42	11 10·5	11 12·3	10 40·0	4·2	3·1	10·2	7·6	16·2	12·0
43	11 10·8	11 12·6	10 40·2	4·3	3·2	10·3	7·6	16·3	12·1
44	11 11·0	11 12·8	10 40·4	4·4	3·3	10·4	7·7	16·4	12·2
45	11 11·3	11 13·1	10 40·7	4·5	3·3	10·5	7·8	16·5	12·2
46	11 11·5	11 13·3	10 40·9	4·6	3·4	10·6	7·9	16·6	12·3
47	11 11·8	11 13·6	10 41·1	4·7	3·5	10·7	7·9	16·7	12·4
48	11 12·0	11 13·8	10 41·4	4·8	3·6	10·8	8·0	16·8	12·5
49	11 12·3	11 14·1	10 41·6	4·9	3·6	10·9	8·1	16·9	12·5
50	11 12·5	11 14·3	10 41·9	5·0	3·7	11·0	8·2	17·0	12·6
51	11 12·8	11 14·6	10 42·1	5·1	3·8	11·1	8·2	17·1	12·7
52	11 13·0	11 14·8	10 42·3	5·2	3·9	11·2	8·3	17·2	12·8
53	11 13·3	11 15·1	10 42·6	5·3	3·9	11·3	8·4	17·3	12·8
54	11 13·5	11 15·3	10 42·8	5·4	4·0	11·4	8·5	17·4	12·9
55	11 13·8	11 15·6	10 43·1	5·5	4·1	11·5	8·5	17·5	13·0
56	11 14·0	11 15·8	10 43·3	5·6	4·1	11·6	8·6	17·6	13·1
57	11 14·3	11 16·1	10 43·5	5·7	4·2	11·7	8·7	17·7	13·1
58	11 14·5	11 16·3	10 43·8	5·8	4·3	11·8	8·8	17·8	13·2
59	11 14·8	11 16·6	10 44·0	5·9	4·4	11·9	8·8	17·9	13·3
60	11 15·0	11 16·8	10 44·3	6·0	4·5	12·0	8·9	18·0	13·4

45	SUN PLANETS	ARIES	MOON	v or d	Corrⁿ	v or d	Corrⁿ	v or d	Corrⁿ
s	° ′	° ′	° ′	′	′	′	′	′	′
00	11 15·0	11 16·8	10 44·3	0·0	0·0	6·0	4·6	12·0	9·1
01	11 15·3	11 17·1	10 44·5	0·1	0·1	6·1	4·6	12·1	9·2
02	11 15·5	11 17·3	10 44·7	0·2	0·2	6·2	4·7	12·2	9·3
03	11 15·8	11 17·6	10 45·0	0·3	0·2	6·3	4·8	12·3	9·3
04	11 16·0	11 17·9	10 45·2	0·4	0·3	6·4	4·9	12·4	9·4
05	11 16·3	11 18·1	10 45·4	0·5	0·4	6·5	4·9	12·5	9·5
06	11 16·5	11 18·4	10 45·7	0·6	0·5	6·6	5·0	12·6	9·6
07	11 16·8	11 18·6	10 45·9	0·7	0·5	6·7	5·1	12·7	9·6
08	11 17·0	11 18·9	10 46·2	0·8	0·6	6·8	5·2	12·8	9·7
09	11 17·3	11 19·1	10 46·4	0·9	0·7	6·9	5·2	12·9	9·8
10	11 17·5	11 19·4	10 46·6	1·0	0·8	7·0	5·3	13·0	9·9
11	11 17·8	11 19·6	10 46·9	1·1	0·8	7·1	5·4	13·1	9·9
12	11 18·0	11 19·9	10 47·1	1·2	0·9	7·2	5·5	13·2	10·0
13	11 18·3	11 20·1	10 47·4	1·3	1·0	7·3	5·5	13·3	10·1
14	11 18·5	11 20·4	10 47·6	1·4	1·1	7·4	5·6	13·4	10·2
15	11 18·8	11 20·6	10 47·8	1·5	1·1	7·5	5·7	13·5	10·2
16	11 19·0	11 20·9	10 48·1	1·6	1·2	7·6	5·8	13·6	10·3
17	11 19·3	11 21·1	10 48·3	1·7	1·3	7·7	5·8	13·7	10·4
18	11 19·5	11 21·4	10 48·5	1·8	1·4	7·8	5·9	13·8	10·5
19	11 19·8	11 21·6	10 48·8	1·9	1·4	7·9	6·0	13·9	10·5
20	11 20·0	11 21·9	10 49·0	2·0	1·5	8·0	6·1	14·0	10·6
21	11 20·3	11 22·1	10 49·3	2·1	1·6	8·1	6·1	14·1	10·7
22	11 20·5	11 22·4	10 49·5	2·2	1·7	8·2	6·2	14·2	10·8
23	11 20·8	11 22·6	10 49·7	2·3	1·7	8·3	6·3	14·3	10·8
24	11 21·0	11 22·9	10 50·0	2·4	1·8	8·4	6·4	14·4	10·9
25	11 21·3	11 23·1	10 50·2	2·5	1·9	8·5	6·4	14·5	11·0
26	11 21·5	11 23·4	10 50·5	2·6	2·0	8·6	6·5	14·6	11·1
27	11 21·8	11 23·6	10 50·7	2·7	2·0	8·7	6·6	14·7	11·1
28	11 22·0	11 23·9	10 50·9	2·8	2·1	8·8	6·7	14·8	11·2
29	11 22·3	11 24·1	10 51·2	2·9	2·2	8·9	6·7	14·9	11·3
30	11 22·5	11 24·4	10 51·4	3·0	2·3	9·0	6·8	15·0	11·4
31	11 22·8	11 24·6	10 51·6	3·1	2·4	9·1	6·9	15·1	11·5
32	11 23·0	11 24·9	10 51·9	3·2	2·4	9·2	7·0	15·2	11·5
33	11 23·3	11 25·1	10 52·1	3·3	2·5	9·3	7·1	15·3	11·6
34	11 23·5	11 25·4	10 52·4	3·4	2·6	9·4	7·1	15·4	11·7
35	11 23·8	11 25·6	10 52·6	3·5	2·7	9·5	7·2	15·5	11·8
36	11 24·0	11 25·9	10 52·8	3·6	2·7	9·6	7·3	15·6	11·8
37	11 24·3	11 26·1	10 53·1	3·7	2·8	9·7	7·4	15·7	11·9
38	11 24·5	11 26·4	10 53·3	3·8	2·9	9·8	7·4	15·8	12·0
39	11 24·8	11 26·6	10 53·6	3·9	3·0	9·9	7·5	15·9	12·1
40	11 25·0	11 26·9	10 53·8	4·0	3·0	10·0	7·6	16·0	12·1
41	11 25·3	11 27·1	10 54·0	4·1	3·1	10·1	7·6	16·1	12·2
42	11 25·5	11 27·4	10 54·3	4·2	3·2	10·2	7·7	16·2	12·3
43	11 25·8	11 27·6	10 54·5	4·3	3·3	10·3	7·8	16·3	12·4
44	11 26·0	11 27·9	10 54·7	4·4	3·3	10·4	7·9	16·4	12·4
45	11 26·3	11 28·1	10 55·0	4·5	3·4	10·5	8·0	16·5	12·5
46	11 26·5	11 28·4	10 55·2	4·6	3·5	10·6	8·0	16·6	12·6
47	11 26·8	11 28·6	10 55·5	4·7	3·6	10·7	8·1	16·7	12·7
48	11 27·0	11 28·9	10 55·7	4·8	3·6	10·8	8·2	16·8	12·7
49	11 27·3	11 29·1	10 55·9	4·9	3·7	10·9	8·3	16·9	12·8
50	11 27·5	11 29·4	10 56·2	5·0	3·8	11·0	8·3	17·0	12·9
51	11 27·8	11 29·6	10 56·4	5·1	3·9	11·1	8·4	17·1	13·0
52	11 28·0	11 29·9	10 56·7	5·2	3·9	11·2	8·5	17·2	13·0
53	11 28·3	11 30·1	10 56·9	5·3	4·0	11·3	8·6	17·3	13·1
54	11 28·5	11 30·4	10 57·1	5·4	4·1	11·4	8·6	17·4	13·2
55	11 28·8	11 30·6	10 57·4	5·5	4·2	11·5	8·7	17·5	13·3
56	11 29·0	11 30·9	10 57·6	5·6	4·2	11·6	8·8	17·6	13·3
57	11 29·3	11 31·1	10 57·9	5·7	4·3	11·7	8·9	17·7	13·4
58	11 29·5	11 31·4	10 58·1	5·8	4·4	11·8	8·9	17·8	13·5
59	11 29·8	11 31·6	10 58·3	5·9	4·5	11·9	9·0	17·9	13·6
60	11 30·0	11 31·9	10 58·6	6·0	4·6	12·0	9·1	18·0	13·7

INDEX

177

Other books published by
Fishing News Books Ltd

Free catalogue available on request

Advances in aquaculture
Advances in fish science and technology
Aquaculture practices in Taiwan
Aquaculture training manual
Atlantic salmon: its future
Better angling with simple science
British freshwater fishes
Business management in fisheries and
   acquaculture
Commercial fishing methods
Control of fish quality
The crayfish
Culture of bivalve molluscs
Echo sounding and sonar for fishing
The edible crab and its fishery in
   British waters
Eel capture, culture, processing and
   marketing
Eel culture
Engineering, economics and
   fisheries management
European inland water fish: a
   multilingual catalogue
FAO catalogue of fishing gear designs
FAO catalogue of small scale fishing
   gear
Fibre ropes for fishing gear
Fish and shellfish farming in
   coastal waters
Fish catching methods of the world
Fisheries oceanography and ecology
Fisheries of Australia
Fisheries sonar
Fishermen's handbook
Fishery development experiences
Fishing boats and their equipment
Fishing boats of the world 1
Fishing boats of the world 2
Fishing boats of the world 3
The fishing cadet's handbook
Fishing ports and markets
Fishing with electricity
Fishing with light
Freezing and irradiation of fish

Freshwater fisheries management
Glossary of UK fishing gear terms
Handbook of trout and salmon diseases
How to make and set nets
Introduction to fishery by-products
The lemon sole
A living from lobsters
Making and managing a trout lake
Marine fisheries ecosystem
Marine pollution and sea life
Marketing in fisheries and aquaculture
Mending of fishing nets
Modern deep sea trawling gear
More Scottish fishing craft and
   their work
Multilingual dictionary of fish
   and fish products
Netting materials for fishing gear
Pair trawling and pair seining
Pelagic and semi-pelagic trawling gear
Penaeid shrimps – their biology
   and management
Planning of aquaculture development
Power transmission and automation
   for ships and submersibles
Refrigeration on fishing vessels
Salmon and trout farming in Norway
Salmon fisheries of Scotland
Scallop and queen fisheries in the
   British Isles
Scallops and the diver-fisherman
Seine fishing
Squid jigging from small boats
Stability and trim of fishing vessels
The stern trawler
Study of the sea
Textbook of fish culture
Training fishermen at sea
Trends in fish utilization
Trout farming handbook
Trout farming manual
Tuna: distribution and migration
Tuna fishing with pole and line